Ulrich Hammes

Robust Positioning Algorithms for Wireless Networks

AF063904

Ulrich Hammes

Robust Positioning Algorithms for Wireless Networks

Statistical Approaches to Non-Line-of-Sight (NLOS) Mitigation

Südwestdeutscher Verlag für Hochschulschriften

Imprint
Any brand names and product names mentioned in this book are subject to trademark, brand or patent protection and are trademarks or registered trademarks of their respective holders. The use of brand names, product names, common names, trade names, product descriptions etc. even without a particular marking in this work is in no way to be construed to mean that such names may be regarded as unrestricted in respect of trademark and brand protection legislation and could thus be used by anyone.

Publisher:
Südwestdeutscher Verlag für Hochschulschriften
is a trademark of
Dodo Books Indian Ocean Ltd., member of the OmniScriptum S.R.L Publishing group
str. A.Russo 15, of. 61, Chisinau-2068, Republic of Moldova Europe
Printed at: see last page
ISBN: 978-3-8381-1571-9

Zugl. / Approved by: Darmstadt, TU Darmstadt, Diss., 2010

Copyright © Ulrich Hammes
Copyright © 2010 Dodo Books Indian Ocean Ltd., member of the OmniScriptum S.R.L Publishing group

Kurzfassung

Die vorliegende Arbeit beschäftigt sich mit der Positionsbestimmung von elektronischen Sendern (z.B. Mobiltelefon) innerhalb drahtloser Netzwerke unter Verwendung von Signalparametern wie dem Einfallswinkel (*Angle-of-Arrival*) oder der Ankunftszeit (*Time-of-Arrival*). Diese Signalparameter werden beispielsweise an den stationären Empfängern des drahtlosen Netzwerks geschätzt.
Wenn eine Sichtverbindung (*Line-of-Sight* (LOS)) zwischen Sender und Empfänger besteht, kann mittels Trilateration oder Triangulation eine hohe Positionierungsgenauigkeit erzielt werden. In der Realität trifft die Annahme einer Sichtverbindung zwischen Sender und Empfänger jedoch selten zu. Durch Hindernisse auf dem Übertragungsweg, wie z.B. Häuser oder Bäume, wird das Signal gegebenenfalls mehrfach reflektiert und erreicht so den Empfänger auf einem indirekten Pfad. Dies wird in der Literatur als *Non-Line-of-Sight* (NLOS)-Ausbreitung bezeichnet und führt bei der Schätzung der oben genannten Signalparameter zu großen Fehlern. Diese Fehler werden hier als statistische Ausreißer modelliert und haben zur Folge, dass herkömmliche Positionierungsverfahren wie der Kleinste-Quadrate-Schätzer (*least-squares estimator*) oder erweiterte Kalman Filter (EKF) zu erheblichen Ungenauigkeiten führen.
Aus diesem Grund werden Verfahren benötigt, die sich robust gegenüber der LOS-Annahme verhalten und auch in schwierigen Umgebungen eine angemessene Genauigkeit erreichen.
Da in der Praxis die statistische Verteilung der NLOS-Ausreißer unbekannt ist, schlagen wir vor, diese Verteilung aus den Beobachtungen heraus nicht-parametrisch zu schätzen. Die geschätzte Verteilung wird innerhalb eines parametrischen Modells verwendet, um die Position eines stationären Senders mit Hilfe des Maximum-Likelihood-Prinzips zu bestimmen. Dieser als semi-parametrisch bezeichnete Ansatz erzeugt eine signifikante Erhöhung der Positionierungsgenauigkeit gegenüber konventionellen Methoden in NLOS-Umgebungen, während in LOS-Umgebungen eine ähnliche Genauigkeit wie der Kleinste-Quadrate-Schätzer erreicht werden kann.
Dieser Ansatz wird innerhalb der Arbeit für einen räumlich nicht-stationären Sender unter Verwendung eines EKF ausgebaut. Dabei werden die Gleichungen des EKF für jeden Zeitpunkt in ein lineares Regressionsmodell umformuliert und der semi-parametrische Schätzer wird verwendet, um die Position und Geschwindigkeit des Senders zu schätzen.
Für das Problem eines räumlich nicht-stationären Senders wird weiterhin ein Zielverfolgungsalgorithmus vorgeschlagen, der einen EKF und eine parametrische, robustifizierte Version desselben parallel verwendet und je nach Situation unterschiedlich stark gewichtet. Dadurch kann eine hohe Positionsgenauigkeit in LOS-Umgebungen sowie

Robustheit gegenüber NLOS-Messungen erzielt werden.

Darüber hinaus stellen wir einen kombinierten NLOS-Erkennungs- und Zielverfolgungsalgorithmus vor. Ein Hypothesentest detektiert dabei Positionsmessungen, die aufgrund von NLOS-Ausreißern fehlerhaft sind. Diese Beobachtungen werden verworfen und die verbleibenden Messungen werden für den Aktualisierungsschritt des Kalman Filters verwendet. Da nicht bekannt ist welche dieser Messungen die höchste Präzision erzielen, werden sie mit unterschiedlichen Wahrscheinlichkeiten gewichtet.

Alle im Rahmen der Arbeit vorgeschlagenen Algorithmen erzielen höhere Positionierungsgenauigkeiten in NLOS-Umgebungen als verschiedene Vergleichsmethoden aus der Literatur. Dabei werden keine Kenntnisse der statistischen Verteilung der NLOS-Ausreißer vorausgesetzt. Eine vergleichbare Genauigkeit zu Standard-Verfahren wie z.B. dem Kleinste-Quadrate-Schätzer und dem EKF kann in LOS-Umgebungen erreicht werden.

Abstract

In this thesis, we consider the problem of finding the geographic position of a transmitter device (e.g. mobile phone), denoted as user equipment (UE), based on signal parameter estimates such as angle-of-arrival or time-of-arrival that are provided by surrounded sensors or base stations.

If line-of-sight (LOS) channels between the UE and the base stations exists, high positioning accuracy can be obtained using trilateration or triangulation techniques. However, this assumption is ideal and not often encountered in practice. Especially in urban areas and hilly terrain, reflections at obstacles such as buildings and trees occur which force the signals of the UE to arrive at the base station via an indirect path. This phenomenon, called non-line-of-sight (NLOS) propagation, leads to erroneous signal parameter estimates that can strongly differ from the true ones and are thus modeled as outliers here. These NLOS errors result in large positioning errors when using standard techniques such as least-squares estimation and extended Kalman filtering. Thus, positioning algorithms that are robust against deviations from the LOS assumption are required.

Since the statistics of the errors due to NLOS propagation are unknown in general we develop estimators that determine the NLOS error statistics from the observations non-parametrically. This estimate is then used in a parametric model to obtain the position estimate of the UE based on the maximum likelihood principle. The approach is termed semi-parametric since non-parametric pdf estimation is used for position estimation within a parametric signal model. A significant improvement in positioning accuracy with respect to conventional techniques is achieved in NLOS environments. For LOS environment, where Gaussian sensor noise is predominant, the proposed approach performs similar to a least-squares estimator.

This approach is further extended to the case when the UE is moving over time. For this purpose, the framework of an extended Kalman filter (EKF) is used where the EKF equations are rewritten into a linear regression model at each time step and the semi-parametric estimator is used to solve for the state vector, containing position and velocity of the UE. Furthermore, a multiple model tracking algorithm is proposed that combines the advantages of robust EKFs and the standard EKF to achieve high accuracy in both LOS and NLOS environments.

Finally, a different approach for positioning of a moving UE in NLOS environments is developed. It is based on a joint outlier detection and tracking algorithm where the errors due to NLOS effects are detected and discarded and the remaining measurements are used for updating the position estimate. Since we do not know which of them yields highest precision the remaining measurements are weighted with different probabilities to obtain the state estimate at each time step.

The developed tracking algorithms outperform various robust competing estimators found in the literature while no knowledge of the NLOS error statistics is required.

Danksagung

Die vorliegende Arbeit entstand im Rahmen meiner Tätigkeit als wissenschaftlicher Mitarbeiter am Fachgebiet Signalverarbeitung des Instituts für Nachrichtentechnik der Technischen Universität Darmstadt.

Die wissenschaftliche Betreuung erfolgte durch Herrn Prof. Dr.-Ing. Abdelhak Zoubir, dem ich an dieser Stelle für seine Unterstützung und zahlreiche fachliche Diskussionen recht herzlich danken möchte. Des Weiteren bedanke ich mich bei Herrn Prof. Dr. Fredrik Gustafsson für die freundliche Übernahme des Korreferats und sein Interesse an meiner Arbeit. Ebenso möchte ich mich bei Frau Prof. Dr.-Ing. Anja Klein sowie den Herren Prof. Dr.-Ing. Rolf Jakoby und Prof. Dr.-Ing. Gerd Balzer für ihre Mitwirkung in der Prüfungskommission bedanken.
Weiterhin gilt mein herzlicher Dank Dr. Eric Wolsztynski, der mir auch nach seiner Zeit als Postdoc am Fachgebiet Signalverarbeitung mit Rat und Tat zur Seite gestanden hat. Meinen ehemaligen Kollegen Dr. Ramon Brcic und Dr. Christopher Brown danke ich für ihre tatkräftige Unterstützung während meiner Einarbeitungszeit und darüber hinaus.
Überdies möchte ich mich bei Carsten Fritsche für die gute Zusammenarbeit sowie für viele fachliche Diskussionen bedanken. Dies gilt ebenso für Marco Moebus, Christian Debes, Philipp Heidenreich, Dominik Müller, Michael Rübsamen, Michael Muma, Raquel Fandos und Weaam Alkhaldi, die in der ein oder anderen Weise zum Gelingen meiner Arbeit beigetragen haben.
Darüber hinaus bedanke ich mich bei allen Kollegen des Fachgebiets Signalverarbeitung für die angenehme Arbeitsatmosphäre sowie bei den Studenten, deren Studien- und Diplomarbeiten ich betreut habe, was mir oft viel Freude bereitet hat.
Schließlich möchte ich mich auch bei meiner Familie und meiner Freundin Daniela für die gute Unterstützung und den Rückhalt bedanken, wodurch ich so manches Problem vergessen konnte und somit neue Energie für die Arbeit gewonnen habe.

Darmstadt, im Januar 2010   Ulrich Hammes

Contents

1 Introduction — 11
   1.1 Motivation and Existing Techniques — 11
   1.2 Objectives and Context of Research — 14
   1.3 Assumptions — 15
   1.4 Contributions — 16
   1.5 Scope and Overview — 17

2 Estimation Techniques for Wireless Positioning — 18
   2.1 Problem Statement for Geolocation — 18
      2.1.1 Signal Model for a Stationary User Equipment — 18
      2.1.2 Maximum Likelihood Estimation — 20
      2.1.3 Principles of Robust Estimation — 21
         2.1.3.1 (Geo)-Location Estimation — 22
         2.1.3.2 Scale Estimation — 26
      2.1.4 Adaptive Estimation — 27
         2.1.4.1 Adaptive Parametric Estimation — 27
         2.1.4.2 Semi-Parametric Estimation — 29
   2.2 Problem Statement for Tracking — 34
      2.2.1 Signal Model for a Moving User Equipment — 35
         2.2.1.1 Nonlinear System Model — 35
         2.2.1.2 Jump-Markov Nonlinear Model — 36
      2.2.2 Nonlinear Filtering — 38
      2.2.3 Hybrid Nonlinear Filtering — 41
      2.2.4 State Estimation in the Presence of Outliers — 43
         2.2.4.1 General Problem — 43
         2.2.4.2 Robust Regression Kalman Filtering — 44

3 Robust Geolocation — 45
   3.1 Problem Statement — 45
      3.1.1 State of the Art — 46
      3.1.2 Linearization — 48
   3.2 Approaches for Position Estimation — 49
      3.2.1 Maximum Likelihood and Least-Squares Estimation — 49
      3.2.2 Robust M-estimation — 51
      3.2.3 Positioning Based on Semi-Parametric Estimation — 53
         3.2.3.1 General Concept — 53
         3.2.3.2 Transformation KDE for Asymmetric Noise Densities — 54

		3.2.3.3	Selection of the Tuning Parameters δ and λ	56
		3.2.3.4	Algorithm	57
3.3	Numerical Study			58
	3.3.1	Simulation Environments and Settings		58
	3.3.2	Simulation Results		61
		3.3.2.1	Impact of λ on the Position Estimates of the Semi-Parametric Estimator	61
		3.3.2.2	NLOS Outliers Modeled as a Shifted Gaussian pdf	63
		3.3.2.3	NLOS Outliers Modeled as an Exponential pdf	65
		3.3.2.4	Comments on Computational Complexity	68
3.4	Discussion and Conclusions			69

4 Robust Tracking — 71

4.1	Problem Statement			71
	4.1.1	Signal Model		71
	4.1.2	State of the Art		73
		4.1.2.1	Extended Kalman Filter	73
		4.1.2.2	Robust Extended Kalman Filter	75
		4.1.2.3	Interacting Multiple Model Algorithm and Existing Approaches	77
		4.1.2.4	Other Existing Techniques for Tracking an UE	79
4.2	Noise-adaptive EKF using Semi-Parametric Estimation			80
	4.2.1	General Idea		80
	4.2.2	Algorithm		81
4.3	Robust Tracking based on M-Estimation and Interacting Multiple Model Algorithm			82
	4.3.1	Model Reduction		82
	4.3.2	Algorithm		83
4.4	Numerical Study			87
	4.4.1	Simulation Environments and Settings		87
		4.4.1.1	Simulation Environments	87
		4.4.1.2	Settings of the Trackers	88
	4.4.2	Simulation Results		89
		4.4.2.1	NLOS Outliers Modeled by a Markov Chain	89
		4.4.2.2	Observations are iid	95
4.5	Discussion and Conclusions			97

5 Tracking based on Outlier Detection and Data Association — 99

5.1	Problem Statement			99
	5.1.1	Signal Model		99

		5.1.2	Context of Research and Existing Methods	100
	5.2	Modified Probabilistic Data Association		101
		5.2.1	General Concept	101
		5.2.2	Grouping and Positioning	102
		5.2.3	Algorithm	103
			5.2.3.1 Kalman Prediction	103
			5.2.3.2 NLOS Detection	103
			5.2.3.3 Data Association	104
			5.2.3.4 Update	106
	5.3	Numerical Study		107
		5.3.1	Simulation Environments and Settings	107
		5.3.2	Simulation Results	109
			5.3.2.1 NLOS Outliers Modeled by a Markov Chain	109
			5.3.2.2 NLOS Outliers Modeled as iid	112
	5.4	Discussion and Conclusions		116

6 Conclusions and Future Work — 118
- 6.1 Conclusions — 118
  - 6.1.1 Stationary User Equipment — 118
  - 6.1.2 Moving User Equipment — 119
- 6.2 Future Work — 122
  - 6.2.1 Stationary User Equipment — 122
  - 6.2.2 Moving User Equipment — 122

Appendix — 124
- A.1 Choice of Clipping Parameters for Geolocation — 124
- A.2 Robust Tracking — 126
  - A.2.1 Choice of Clipping Parameters — 126
  - A.2.2 Transition Probabilities of Reduced Model — 127
  - A.2.3 Markov Matrices — 128

List of Acronyms — 129

List of Symbols — 131

Bibliography — 137

Publications — 145

Chapter 1
Introduction

This thesis deals with the problem of determining the geographic position of a radio transmitter by exploiting signal parameters such as time-of-arrival (TOA), angle-of-arrival (AOA) or others together with the geometry of a network of receivers. If direct physical connections between the transmitter and each of the receivers exist, meaning the channel is in line-of-sight (LOS), accurate position estimates can be obtained by trilateration or triangulation techniques. However, in practice obstacles such as buildings and trees reflect the signals and hinder them to arrive at the receivers via the direct path. This phenomenon, called non-line-of-sight (NLOS) propagation, yields erroneous signal parameters and consequently high positioning errors. The goal of this thesis is to design statistical algorithms that achieve similar positioning accuracy to conventional techniques in LOS environments and do not degrade significantly when the degree of NLOS propagation increases.

1.1 Motivation and Existing Techniques

Finding the geographic location of a user equipment (UE) (object or human being with a transmitter device) is an important task in many civilian and military applications such as emergency services, yellow page services, intelligent transport systems among others [13, 37, 41, 89, 102]. For this purpose electronic navigation systems that are categorized into Global Navigation Satellite System (GNSS) [76] and network-based positioning systems [41, 89] can be used. The former ones such as Global Positioning System (GPS) or *Galileo*, use an electronic device to receive radio signals transmitted from satellites. These signals are processed to yield the position of the UE. They provide positioning accuracy up to a few meters given that no obstacles block the signal waveforms from the satellites [76]. When it comes to worse conditions like indoor environments, harsh urban environments or hilly terrains, GNSSs become unreliable because obstacles in the propagation path lead to strong signal degradation and consequently to erroneous position estimates. In these environments network-based positioning systems use network infrastructure to support or replace GNSSs. Supporting GNSS position estimates with network-based approaches is termed *hybrid positioning* [34, 74]. However, these techniques require more sophisticated hardware and are therefore of limited use in many applications.

Instead we concentrate on systems for which positioning completely relies on the network infrastructure and the transmitted signals without using any GNSS signals. Theses systems include cellular network infrastructures like Global System for Mobile Communications (GSM) or Universal Mobile Telecommunications System (UMTS) [13] as well as sensor networks [56]. Based on any network one can use either geometric [19, 41, 89, 100] or mapping approaches [30, 66, 96] to determine the position of the UE.

Mapping approaches, also known as *location fingerprinting* [30, 66, 96], are based on correlation between local measurements from the UE, such as received-signal-strength (RSS), with measurements stored in a database which need to be recorded beforehand. The database contains a grid of geometric positions assigned to local measurements. Then, the best match between the local measurements obtained by the UE and one of the measurements in the database yields the position of the UE. However, estimation accuracy of mapping approaches is limited by the grid resolution in the database and the measurement accuracy of the UE. Furthermore, frequent and extensive measurement campaigns are required to accommodate the time varying characteristic of wireless channels and upgrades of the network infrastructure.

In contrast, geometric approaches, which use measurements and the geometry of the network to infer the UE position, can usually achieve higher accuracy than mapping approaches when a direct LOS channel between the UE and each fixed terminal (FT) exists. The FT can be a base station in a cellular network or a node in a sensor network. Direct geometric approaches [108] use the signals traveling between the nodes for positioning. On the other hand, two step approaches first estimate the signal parameters such as TOA, AOA, time-difference-of-arrival (TDOA) or RSS. These signal parameters can be estimated at each FT based on the received signal waveforms transmitted from the UE and are used for positioning together with the geometry of the FTs. Algorithms for estimating the signal parameters such as TOA and AOA are treated in [17, 63] and are out of the scope of this work. Depending on the system, the position of the UE can be estimated by the UE itself, or it can be estimated by a central processor that obtains the signal parameters via the FTs [41].

Figure 1.1 illustrates an example for wireless positioning based on TOA and AOA parameter estimates, respectively. We assume that a direct physical connection between the UE and each of the FTs exists, meaning the channel is in LOS. For TOA positioning, illustrated in Figure 1.1(a), the travel time of the signal from the UE to each FT is measured and the distances between the UE and the FTs are obtained by multiplying the TOA estimates by the speed of light. Then, the position of the UE is the intercept of the three circles also known as trilateration. Note that at least three FTs are required to obtain a unique solution. In contrast, usually two FTs are required for positioning based on AOA parameter estimates, which is depicted in Figure 1.1(b).

1.1 Motivation and Existing Techniques

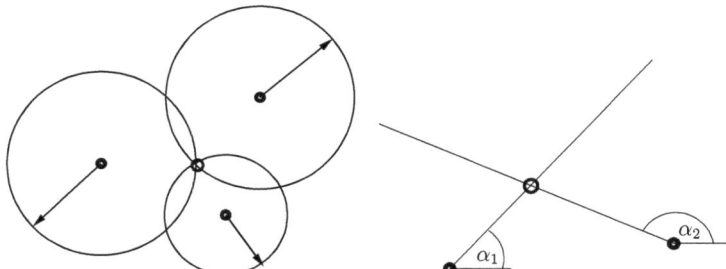

(a) Trilateration based on TOA estimates. (b) Triangulation based on AOA estimates.

Figure 1.1: Geometric approaches for localization of a transmitter device.

The angle or direction of the received waveform impinging on an antenna array or a set of fixed directional antennas is estimated for each FT. Based on these AOA estimates the position of the UE is obtained by finding the intercept of the two lines, known as triangulation. In the presence of sensor noise small positioning errors occur which can be reduced further when motion models are incorporated for tracking position estimates over time.

However, if the assumption of an LOS channel for any of the FTs is not fulfilled, the estimated signal parameters can completely differ from the ones expected under the LOS assumption. This phenomenon known as NLOS propagation is illustrated in Figure 1.2 where a direct LOS channel exists between one FT and the UE and the other FT suffers from an indirect path or NLOS channel. In reality, obstacles such as buildings, hills and trees in outdoor environments and furniture and walls in indoor environments often prevent the signals from propagating along the direct path and reflection or diffraction occur at these obstacles. Thus, the assumption of an LOS channel for every FT is ideal and not often encountered in practice [95, 111].

The errors introduced by NLOS propagation into the signal parameters are different for each scenario which make a deterministic description impractical. Instead, describing these errors stochastically better suits the phenomenon. Since the errors due to NLOS propagation are very large with respect to the errors due to sensor noise, we model the NLOS errors as statistical outliers[1] here.

For simplicity many standard algorithms do not take into account NLOS propagation and assume an LOS channel for every FT [13, 22, 89, 111]. These algorithms lose significantly in positioning accuracy when NLOS propagation occurs, thus algorithms that are robust against NLOS errors are required. The NLOS problem is considered as one of the most severe problems in wireless positioning [13, 111].

---
[1]An outlier is an observation that is numerically distant from the rest of the data.

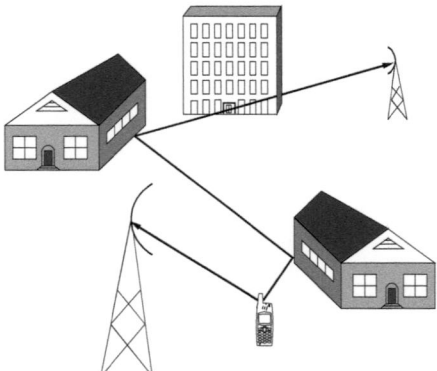

Figure 1.2: Urban scenario with an LOS and an NLOS channel.

1.2 Objectives and Context of Research

The goal of this thesis is to develop positioning algorithms, based on statistical tools from estimation theory, that perform close to standard techniques in LOS environments and do not lose much in positioning accuracy in NLOS scenarios, when the percentage of NLOS outliers increases.

Traditional parametric approaches for positioning or NLOS detection, that are treated in [10, 67, 83], require knowledge of the NLOS error statistics which is usually not available in practice. If the underlying model differs from reality, a model mismatch and consequently high positioning errors are expected.

Instead two approaches from the statistical literature are deployed here to obtain high precision in LOS and robustness in NLOS environments:

- *Robust Statistics* [49,51,86] describe estimators that are robust to some deviation from a presumed model. We can assume a parametric model (e.g. the Gaussian distribution) for sensor noise under LOS propagation and design an estimator that is robust when less than 50% of the observations deviate from the presumed model. However, the prize we pay for decreasing the positioning errors in NLOS environments is the loss of positioning accuracy in the nominal case, i.e., LOS. Thus, there is always a trade-off between robustness and efficiency.

  Positioning algorithms based on *robust statistics* [51] that maintain robustness over a certain class of different models are proposed in [18,72,80,97] to determine the position of a UE.

- *Semi-parametric Statistics* [8] are used to estimate the model non-parametrically. Based on this estimate the position of the UE can be determined. This approach has the advantage that it adapts to any scenario, since less assumptions are made and the noise statistics are estimated for every set of observations.
  Semi-parametric approaches have been addressed in [36] for NLOS identification and in [73, 98] for NLOS mitigation and positioning of a stationary UE.

Another class of algorithms, suggested in [21, 22, 111], are distribution-free, meaning they do not assume any NLOS error model and calculate the position estimates based on some criterion.

While the above mentioned statistical approaches take into account all measurements an alternative is to perform NLOS outlier detection and rejection and process the remaining observations with conventional algorithms. For this purpose robust outlier detection can be used [86].

The focus of this work lies in achieving accurate position estimates in mixed LOS/NLOS environments while keeping computational complexity on a reasonable level.

1.3 Assumptions

In the remainder of this thesis, apart from Chapter 2, we concentrate on TOA based positioning using electromagnetic signals. Measuring the TOA or time delay between the FTs and the UE requires synchronization of the clocks of all devices which is impractical. Instead, one can use measures such as timing advance (TA) in GSM [92], round trip time (RTT) in UMTS [14] or round trip time of flight (RTOF) in indoor logistic systems [102] that measure the time of flight of the signal from an FT to the UE and back. The TOA estimate is then half the time of flight. To determine this quantity, time delay estimation techniques [17], such as correlation techniques, are required which are beyond the scope of this work. The Euclidean distances between the UE and the FTs describe the direct propagation paths under the assumption of LOS. Since any indirect path is always longer than the direct one, shadowing due to NLOS results in a longer propagation path leading to positively biased error statistics [13, 92] under certain assumptions[2]. Throughout the remaining part of this thesis we assume that the NLOS error statistics for TOA range measurements are positively biased.

---

[2]The TOA error statistics depends on many parameters such as channel conditions, signal bandwidth, receiver hardware and others. Here, without loss of generality, we consider systems whose NLOS error statistics have a positive mean, e.g., in [95] it is found that the NLOS error statistics are positively biased in a GSM network.

We use the term "noise" for both, sensor noise and perturbations/interferences due to NLOS propagation. Furthermore, we assume throughout this work that all quantities associated with the NLOS error statistics are unknown.

The techniques presented here can be used for both 2D and 3D positioning. For ease of explanation we focus on 2D positioning throughout this thesis.

1.4 Contributions

- Semi-parametric estimator for determining the position of a stationary UE based on TOA measurements [43, 45]. The degree of asymmetry of the NLOS error statistics is determined and used to iteratively calculate the NLOS error statistics together with the position of the UE.

- Semi-parametric estimator for parameter estimation that is able to cope with outliers following a symmetric distribution [44, 46]. The influence of outliers is estimated and used to determine the noise statistics and the parameter of interest in an iterative way.

- Semi-parametric tracker for positioning of a moving UE based on TOA measurements. The impact of positive outliers is mitigated by estimating their statistics which are used for position estimation [45].

- Robust multiple model tracker for positioning of a moving UE which uses two filters in parallel [47]. The conventional one is well suited for LOS environments whereas a robust filter better suits NLOS environments. Each of them provide position estimates that are combined based on their confidence to yield the final position estimate. Consequently, the multiple model framework circumvents the trade-off between efficiency in the nominal case and robustness in NLOS.

- A combined NLOS outlier detection and tracking approach for positioning of a moving UE. Different subgroups of TOA measurements are constructed to obtain different position estimates. Erroneous estimates are detected and discarded whereas the accepted ones are assigned with different probabilities to yield the final position estimate [48].

1.5 Scope and Overview

Chapter 2 covers the theory of classical, robust and semi-parametric estimation in the context of positioning based on TOA, AOA, TDOA or RSS and is required for the remainder of the thesis. The signal models for a stationary and a moving UE are introduced and optimal, suboptimal and robust estimation schemes are explained. In addition, a semi-parametric estimator is presented [44] that assumes a symmetric error distribution.

In Chapter 3 and 4 the symmetric assumption is relaxed and robust parametric and semi-parametric estimators are developed that perform well in noise environments in which outliers follow an asymmetric error statistics. Chapter 3 covers positioning for a stationary UE where a semi-parametric positioning algorithm is developed. Since the degree of asymmetry of the NLOS errors is unknown it is estimated and incorporated into the semi-parametric estimators [43, 45].

Chapter 4 deals with positioning of a moving UE where two robust tracking algorithms are developed [45, 47]. The first one is based on the semi-parametric approach proposed in Chapter 3 and the second one uses a multiple model approach where conventional and robust estimators are used in parallel to accommodate changing LOS/NLOS conditions. Chapter 5 presents a different approach for tracking a moving UE [48]. First the NLOS outliers are detected and discarded whereas the remaining observations are weighted with different probabilities.

Conclusions are drawn in Chapter 6 and an outlook for future work is presented.

Chapter 2
Estimation Techniques for Wireless Positioning

In this section estimation techniques for determining the position of a UE based on signal parameters such as TOA, AOA, TDOA or RSS are presented. The chapter is divided into two parts: The first part deals with estimating the position of a stationary terminal. Classical optimal and robust estimation techniques are discussed and semi-parametric approaches are introduced. Estimators based on semi-parametric estimation are presented [44, 46].

The second part concerns estimating the position of a moving UE in a Bayesian framework where two different measurement models are presented. The optimal solution is briefly described and suboptimal approaches are sketched. At the end an overview of robust state estimators is given.

2.1 Problem Statement for Geolocation

In this section, estimation techniques for a stationary UE, also known as geolocation, are discussed where it is assumed that signal parameters such as TOA, AOA, TDOA or RSS are collected and batch-processing is performed to estimate the parameter of interest $\boldsymbol{\theta} = [x\ y]^\mathsf{T}$, i.e., the x- and y-position of the UE. Here, no prior knowledge about $\boldsymbol{\theta}$ is available and it is assumed to be deterministic.

2.1.1 Signal Model for a Stationary User Equipment

Consider a UE located near to M FTs. Then, the received signal $\mathbf{y} = [y_1, \ldots, y_M]^\mathsf{T} \in \mathbb{R}^M$ is

$$\mathbf{y} = \mathbf{h}(\boldsymbol{\theta}) + \mathbf{v}, \tag{2.1}$$

where $\mathbf{h}(\boldsymbol{\theta}) = [h_1(\boldsymbol{\theta}), \ldots, h_M(\boldsymbol{\theta})]^\mathsf{T}$ is a real, nonlinear vector function $\mathbb{R}^2 \to \mathbb{R}^M$ describing the relationship between $\boldsymbol{\theta}$ and the FT positions $x_{\mathsf{FT},m}$, $y_{\mathsf{FT},m}$, $m = 1, \ldots, M$. An overview for different signal parameters can be found in Table 2.1. In LOS environments when sensor noise is predominant, the random variables in vector $\mathbf{v} = [v_1, \ldots, v_M]^\mathsf{T}$ are assumed independent and identically distributed (iid) zero-mean Gaussian.

2.1 Problem Statement for Geolocation

Signal parameter	$h(\boldsymbol{\theta})$
TOA	$\sqrt{(x - x_{\text{FT},m})^2 + (y - y_{\text{FT},m})^2}$
AOA	$\tan^{-1}\left(\frac{y - y_{\text{FT},m}}{x - x_{\text{FT},m}}\right)$
TDOA	$\sqrt{(x - x_{\text{FT},m})^2 + (y - y_{\text{FT},m})^2} - \sqrt{(x - x_{\text{FT},1})^2 + (y - y_{\text{FT},1})^2}$

Table 2.1: Relationship between different signal parameters, the position of the FTs and the position of the UE $\boldsymbol{\theta} = [x\ y]^{\mathsf{T}}$.

However, when NLOS propagation occurs, the signal is reflected and diffracted by several obstacles and does not travel on the direct path anymore. This results in a significant change of the signal parameters with respect to the signal parameters under the LOS assumption. Examples for these changes are, an increase in TOA, since the indirect path is longer than the direct one [41], a drop in RSS due to stronger signal attenuation and that the AOA of an indirect path can differ completely from the AOA of the direct path. These unforeseen events can lead to large errors in the measured signal parameters and are therefore modeled as measurement outliers here. The probability that NLOS propagation and consequently outliers occur is ε. Then, the class $\mathcal{F}$ of the overall error distribution describing LOS and NLOS environments is defined as an ε-contaminated mixture model [51], i.e.,

$$\{\mathcal{F} : f_V(v) = (1 - \varepsilon)\mathcal{N}(v; 0, \sigma_G^2) + \varepsilon\mathcal{H}(v)\} \tag{2.2}$$

where $0 \leq \varepsilon \leq 1$, $\mathcal{N}(v; 0, \sigma_G^2)$ is a zero-mean normal probability density function (pdf) with variance σ_G^2 modeling sensor noise and $\mathcal{H}(v)$ is the convolution of the pdf f_η, modeling the errors due to NLOS propagation, and the Gaussian density $\mathcal{N}(v; 0, \sigma_G^2)$. Note that the variance of the measurement noise is $\sigma_V^2 = (1 - \varepsilon)\sigma_G^2 + \varepsilon(\sigma_G^2 + \sigma_\eta^2)$ and increases with the degree of NLOS propagation ε, resulting in lower signal-to-noise-ratios (SNRs).

Different models for the NLOS error depending on the estimated signal parameters, e.g., a uniform distribution for AOA and a shifted Gaussian or an exponential distribution for TOA can be found in the literature [15,41,73]. Which error model for a specific signal parameter best fits reality depends on many different things such as system parameters and channel characteristics and may vary for different environments. To avoid this limitation $\mathcal{H}(v)$ is left unspecified for a specific signal parameter and algorithms are required that perform well over a certain class of error pdfs. This kind of problem can be tackled using the theory of robust statistics [49,51] and semi-parametric statistics [8]. It is desired that the algorithms perform similar to conventional ones in LOS environments and are robust against outliers due to the NLOS effect when ε increases. In this thesis, robust positioning algorithms based on the above mentioned theories and assumptions are developed.

2.1.2 Maximum Likelihood Estimation

We want to estimate $\boldsymbol{\theta}$ from the measurements $\mathbf{y}$ and the known positions of M FTs. The estimator that maximizes the likelihood for a parameter $\boldsymbol{\theta}$ given a sample of observations y_m, $m = 1, \ldots, M$ from (2.1) is called a maximum likelihood estimator (MLE). When $f_V(v)$ from (2.2) is known, the likelihood function is

$$L(\boldsymbol{\theta}|y_1, \ldots, y_M) = \prod_{m=1}^M f_{Y_m}(y_m|\boldsymbol{\theta}) = \prod_{m=1}^M f_V(y_m - h_m(\boldsymbol{\theta})). \tag{2.3}$$

Then, the maximum likelihood (ML) solution coincides with the minimum of the negative log-likelihood function

$$\hat{\boldsymbol{\theta}}_{\mathsf{ML}} = \arg\min_{\boldsymbol{\theta}} \sum_{m=1}^M -\log f_V(y_m - h_m(\boldsymbol{\theta})). \tag{2.4}$$

Differentiating (2.4) with respect to $\boldsymbol{\theta}$ yields the estimator $\hat{\boldsymbol{\theta}}_{\mathsf{ML}}$ which is the solution of

$$\sum_{m=1}^M \varphi(y_m - h_m(\boldsymbol{\theta})) \frac{\partial h_m(\boldsymbol{\theta})}{\partial \boldsymbol{\theta}} = \mathbf{0}, \tag{2.5}$$

where $\varphi(v) = -f'_V(v)/f_V(v)$, with $f'_V(v) = d f_V(v)/dv$, is the location score function. Note that unless the noise is zero-mean Gaussian which is only true for LOS, the ML solution is nonlinear. If $\varepsilon = 0$, $\varphi(v) = v/\sigma_G^2$ and (2.5) reduces to least-squares estimation.

An estimator is unbiased if

$$\mathsf{E}\{\hat{\boldsymbol{\theta}}\} = \boldsymbol{\theta} \quad \forall\, \boldsymbol{\theta}. \tag{2.6}$$

The MLE is unbiased and optimal for large enough data records, i.e., asymptotically ($M \to \infty$) it achieves the Cramér-Rao Lower Bound (CRLB), a lower bound for the best achievable precision (in terms of variance) for any unbiased estimator [60]. Unbiased estimators that attain the CRLB are termed efficient.

If the pdf $f_V(v)$ satisfies certain regularity conditions[1] the asymptotic covariance of any unbiased estimator $\hat{\boldsymbol{\theta}}$ is greater than or equal to the CRLB, i.e.,

$$\mathsf{var}(\hat{\theta}_i) \geq [\mathbf{J}^{-1}(\boldsymbol{\theta})]_{ii}, \tag{2.7}$$

where $\mathbf{J}(\boldsymbol{\theta})$ is termed the Fisher information matrix. It is defined as

$$[\mathbf{J}(\boldsymbol{\theta})]_{ij} = -\mathsf{E}\left[\frac{\partial^2 \log L(\boldsymbol{\theta}|\mathbf{y})}{\partial \theta_i \partial \theta_j}\right], \quad i,j = 1, \ldots, \dim(\boldsymbol{\theta}), \tag{2.8}$$

where $\dim(\boldsymbol{\theta})$ denotes the dimension of vector $\boldsymbol{\theta}$. The Fisher information matrix is evaluated at the true value of the unknown parameter $\boldsymbol{\theta}$. Calculation of the CRLB

---

[1] $\mathsf{E}\{\frac{\partial \log f_V(\mathbf{y}|\boldsymbol{\theta})}{\partial \boldsymbol{\theta}}\} = \mathbf{0} \quad \forall\, \boldsymbol{\theta}$ [60].

for the position of a UE, based on different signal parameters, can be found in the literature, see [81, 82] and references therein where a Gaussian NLOS error pdf is often assumed for simplicity. Here, we consider Gaussian and non-Gaussian NLOS distributions to better model reality. In the latter case, a closed-form solution for the CRLB for $\varepsilon \neq 0$ usually does not exist and the CRLB can be calculated numerically. If an efficient estimator exists it coincides with the MLE, i.e., asymptotically

$$\hat{\boldsymbol{\theta}}_{\mathsf{ML}} \sim \mathcal{N}(\boldsymbol{\theta}, \mathbf{J}^{-1}(\boldsymbol{\theta})). \tag{2.9}$$

Furthermore, MLEs are consistent meaning they fulfill

$$\lim_{M \to \infty} \Pr[|\hat{\boldsymbol{\theta}}_{\mathsf{ML}} - \boldsymbol{\theta}| > \vartheta] \to 0, \tag{2.10}$$

for every $\vartheta > 0$. Note that perfect knowledge of the pdf $f_V(v)$ including ε and $\mathcal{H}(v)$ with all parameters is required to design an MLE. An alternative is to assume the nominal Gaussian case where $\varepsilon = 0$ and design an ML estimator under the Gaussian assumption, i.e., a least-squares estimator, that is popular for many reasons [61].

However, in practice neither ε nor the contamination density $\mathcal{H}(v)$ are known which makes it impossible to obtain the MLE. On the one hand, assuming a parametric model $\mathcal{H}(v)$ and estimating ε and other parameters results in large estimation errors when the model does not match reality. On the other hand, least-squares estimation suffers from an increase in the position root mean square error (RMSE) when ε increases because outliers are weighted linearly in (2.5) when using the Gaussian score function v/σ_V^2. Therefore, robust estimation techniques are required, i.e., techniques that are close to optimal when $\varepsilon = 0$ and do not degrade in performance significantly when ε increases (up to a certain percentage), meaning when outliers due to the NLOS effect occur.

2.1.3 Principles of Robust Estimation

In LOS propagation which occurs with probability $1 - \varepsilon$ the error statistics in (2.2) are Gaussian. Since the physical environment of the channel completely changes if an obstacle prevents the signal from traveling on the direct path between the FT and UE, the error statistics change as well and are likely to be biased [111] or heavy-tailed[2] [73]. This results in measurement outliers, i.e., data that are far apart from the nominal Gaussian distribution. Thus, robust estimation techniques are required to mitigate the effect of outliers and deviations from the Gaussian (LOS) assumption. Two different theoretical frameworks dealing with this problem are available in the statistics literature and are applied to the NLOS problem in this thesis: The first one,

---
[2]engineers often use the term "impulsive noise" instead.

robust statistics [49, 51, 86] describes how to design parametric estimators that are insensitive to some deviations from the nominal assumption - the LOS assumption in our case. In contrast, the theory of semi-parametric statistics [8] suggests using the measurements of the parametric measurement model (2.1) to estimate the error pdf $f_V(v)$ non-parametrically. Based on this pdf estimate the ML principle can be applied, leading to a noise-adaptive estimator. In the signal processing literature both approaches are termed "robust" since they are stable over a class of different models. A large amount of work has been done in the field of robust signal processing in the context of dealing with outliers/heavy-tailed noise [59]. Applications arise in multiuser detection (MUD) [105, 113], spectral estimation [44], image processing [65], audio signal processing [38] and geolocation [18, 72, 97] among others.

An alternative approach to cope with the NLOS effect is to perform outlier detection [86], discard the detected measurements and use conventional least-squares estimators or techniques based on missing observations [107] for determining the position of the UE. An overview of NLOS detection schemes is given in [91] and references therein. In this thesis, NLOS detection and outlier rejection is investigated for a moving UE in Chapter 5.

2.1.3.1 (Geo)-Location Estimation

The most common way to cope with outliers is to use M-estimation meaning ML type estimation where the likelihood function from (2.3) and (2.4) is replaced by a certain penalty function $\rho(v)$ [49, 51] which is designed such that large residuals due to NLOS perturbations lose their deleterious impact on the parameter estimate $\hat{\boldsymbol{\theta}}$. The M-estimate is given as $\boldsymbol{\theta}$ that minimizes the cost function

$$\hat{\boldsymbol{\theta}} = \arg\min_{\boldsymbol{\theta}} \sum_{m=1}^{M} \rho(y_m - h_m(\boldsymbol{\theta})), \qquad (2.11)$$

assuming that $\rho(v)$ is differentiable we obtain

$$\sum_{m=1}^{M} \psi(y_m - h_m(\boldsymbol{\theta})) \frac{\partial h_m(\boldsymbol{\theta})}{\partial \boldsymbol{\theta}} = \mathbf{0}, \qquad (2.12)$$

where $\psi(v) = \rho'(v)$ is a bounded, odd-symmetric score function. For a symmetric $\mathcal{H}(v)$, consistency (2.10) and asymptotic normality of the estimator based on $\psi(v)$ is ensured by taking an odd function $\psi(v)$, i.e., $\mathsf{E}\{\psi(v)\} = 0$. The asymptotic variance

2.1 Problem Statement for Geolocation

AV of the robust parameter estimate $\hat{\boldsymbol{\theta}}$ for an arbitrary $\psi(v)$ [32, 51] is

$$\mathsf{AV}(\psi, f_V(v)) = \frac{\int_{-\infty}^{\infty} \psi^2(v) f_V(v) dv}{\left(\int_{-\infty}^{\infty} \psi'(v) f_V(v) dv\right)^2} \sum_{m=1}^{M} \frac{\partial h_m(\boldsymbol{\theta})}{\partial \boldsymbol{\theta}} \left(\frac{\partial h_m(\boldsymbol{\theta})}{\partial \boldsymbol{\theta}}\right)^{\mathsf{T}}, \qquad (2.13)$$

where the first factor depends on the chosen score function $\psi(v)$ and the model $f_V(v)$ and the second factor only depends on the deterministic model $h_m(\boldsymbol{\theta})$. Huber [51, 52] proposed a robust minimax location estimator that minimizes the maximum asymptotic variance of the least favorable distribution, which has a Gaussian shape in a certain region around zero and is decaying exponentially beyond this region. Minimizing the maximum asymptotic variance yields

$$\psi_{c_1}(v) = \arg\min_{\psi(v)} \max_{f_V \in \mathcal{F}} \mathsf{AV}(\psi(v), f_V(v)) \qquad (2.14)$$

where the least favorable distribution $f_V(v)$ is chosen from the class of mixture distributions $\mathcal{F}$ in (2.2) to yield $\psi_{c_1}(v)$. The solution of (2.14), meaning the MLE of the least favorable distribution, turns out to be the soft-limiter [51]

$$\psi_{c_1}(v) = \begin{cases} \frac{v}{\sigma_G^2}, & |v| \leq c_1 \sigma_G^2 \\ c_1 \operatorname{sign}(v), & |v| > c_1 \sigma_G^2, \end{cases} \qquad (2.15)$$

where $\operatorname{sign}(v) = v/|v|$ is the signum function and the penalty function corresponding to (2.15) is

$$\rho_{c_1}(v) = \begin{cases} \frac{v^2}{2\sigma_G^2}, & |v| \leq c_1 \sigma_G^2 \\ \frac{v^2 c_1^2}{2} - c_1 |v|, & |v| > c_1 \sigma_G^2, \end{cases} \qquad (2.16)$$

The optimal c_1 in the minimax sense is determined in terms of the noise scale of the nominal distribution σ_G and ε by solving $2\mathcal{N}(c_1\sigma_G; 0, 1)/(c_1\sigma_G) - 2\Phi(-c_1\sigma_G) = \varepsilon/(1-\varepsilon)$ [51] where $\Phi(\cdot)$ is the standard normal cumulative distribution function (cdf). Since ε is unknown in practice, the clipping point can be approximated by a rule of thumb in terms of the noise scale σ_V [105].

However, in some environments such as positioning based on TOA estimates in GSM networks, the NLOS errors are not symmetric around zero [92] resulting in a pdf $\mathcal{H}(v)$ with positive mean. Consequently, Huber's soft-limiter results in biased position estimates since the symmetric assumption is violated. However, in general it achieves a lower positioning RMSE than least-squares approaches.

Huber's work is extended in [24] and an estimator based on the minimax principle (2.14) is derived under the class $\mathcal{F}$ in (2.2) where no symmetry constraint is put on $\mathcal{H}(v)$. The obtained estimator is based on a redescending score function $\psi_{c_1,c_2}(v)$, meaning a score function which decreases to zero beyond a second threshold c_2 and thus rejects large outliers completely. This estimator leads to consistent location estimates when the clipping points are chosen appropriately.

Redescending score functions are suitable for aysmmetric contamination and for cases where one expects a high number of outliers in the data [51]. However, the problem with redescending score functions is that the solution of (2.12) is not unique [24]. Furthermore, in practice, wrongly chosen clipping points often result in convergence problems when using an optimization algorithm for solving (2.12) leading to higher RMSEs [51]. Note that other score functions are available in the literature [49] with one or more clipping points that have to be selected beforehand.

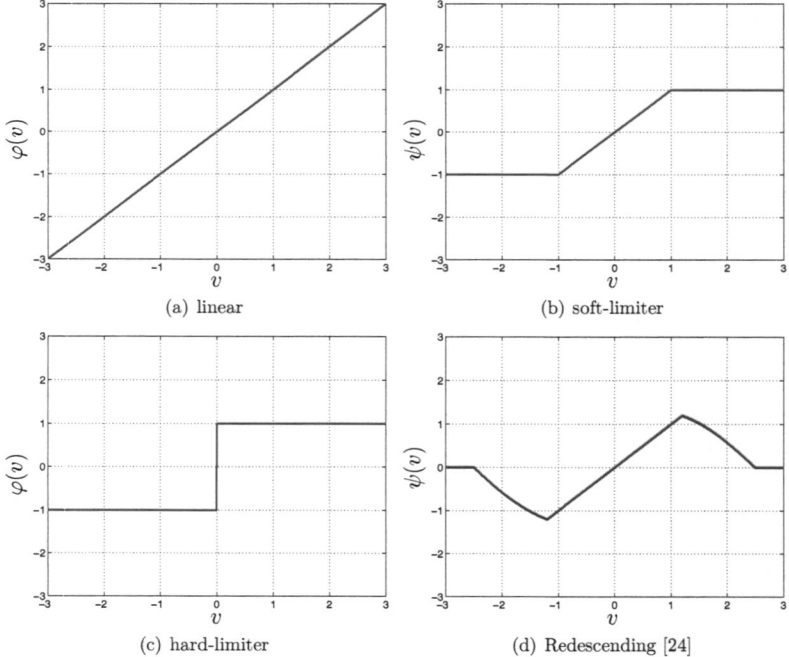

Figure 2.1: Some typical score functions used for parameter estimation. Figure (a) illustrates the ML solution for Gaussian noise and Figure (b) the robust minimax or ML solution for the least favorable symmetric distribution $\mathcal{H}(v)$. Figure (c) depicts the ML solution for Laplacian noise and Figure (d) the robust minimax solution when $\mathcal{H}(v)$ can also be asymmetric. It is assumed that $\sigma_V = 1$ for all functions.

Figure 2.1 shows some typical score functions used for conventional and robust estimation. In order to explain the behavior of the different score functions in LOS and NLOS environments the breakdown point [49] can be used as a measure to quantify robustness. It is defined as the smallest fraction of outliers that may cause the estimator to break down [51]. Parametric score functions functions possess a breakdown point between 0 and 50%.

The ML score function for Gaussian noise is linear and depicted in Figure 2.1(a). Thus, all measurements are weighted linearly in the estimation equation (2.12) resulting in erroneous position estimates when outliers due to the NLOS effect occur. Note that the breakdown point equals to 0 since an arbitrary large outlier can lead to an unbounded parameter estimate $\hat{\boldsymbol{\theta}}$. In contrast the hard-limiter illustrated in Figure 2.1(c) gives equal weight to all measurements and consequently decreases the effect of erroneous data due to NLOS effects. It is the ML score function for Laplacian noise and has a breakdown point of 50%. The estimator in (2.12) based on the hard-limiter (refer to Figure 2.1(c)) minimizes the L1-norm and reduces to the *median* if $\boldsymbol{\theta}$ has dimension one. Even though the hard-limiter has the highest breakdown point among all robust estimators it suffers from a high efficiency loss in the Gaussian case (LOS) due to the discontinuity around zero. Thus, it is of limited use for location estimation which will be shown in Chapter 3. The soft-limiter in 2.1(b) is more appropriate for many applications because it is a compromise between the hard-limiter and the linear least-squares solution achieving higher efficiency in the Gaussian case. The breakdown point depends on the chosen clipping point and the smaller c_1 the higher the breakdown point. The three score functions in Figure 2.1, discussed so far, are based on the assumption of a symmetric $\mathcal{H}(v)$. If this assumption does not hold, e.g., in TOA based positioning the NLOS effect produces a positive bias, estimators based on these score functions result in biased position estimates. The redescending score function depicted in Figure 2.1(d) [24] has two different clipping parameters and discards large outliers completely leading to unbiased estimates in (2.12) when $\mathcal{H}(v)$ is asymmetric given that the clipping points are chosen appropriately.

The clipping parameters for the soft-limiter and the redescending score function need to be adjusted in terms of the noise scale, a nuisance parameter which has to be estimated beforehand or in parallel together with $\boldsymbol{\theta}$. Wrongly chosen clipping parameters can lead to erroneous estimates. If they are chosen too large, almost all measurements including NLOS outliers are weighted linearly leading to a non-robust estimator. In contrast, if the clipping parameters are chosen too small, useful data are discarded leading to a higher variance of the position estimates. A short discussion on robust scale estimation can be found in Section 2.1.3.2.

However, all parametric score functions share the same shortcoming that they do not incorporate the noise pdf $f_V(v)$ into the estimator and can only slightly adapt their shape to the noise by tuning their clipping parameters. For this reason, conventional robust estimation techniques are suboptimal in the Fisher sense for any other noise distribution than the specified one. Instead, robust noise-adaptive approaches, which approximate the true underlying noise pdf $f_V(v)$ parametrically or non-parametrically, are considered in Section 2.1.4.

2.1.3.2 Scale Estimation

Since the variance of the errors of the signal parameter estimates increases with the degree of NLOS contamination, we have to estimate the standard deviation or scale of the noise to incorporate it in the bounded score function. Taking the standard deviation, i.e.,

$$\hat{\sigma}_V = \sqrt{\frac{1}{M-1} \sum_{m=1}^{M} \left(v_m - \frac{1}{M} \sum_{m=1}^{M} v_m\right)^2} \quad (2.17)$$

fails if the noise samples are contaminated by NLOS outliers since they are squared in (2.17) and can have an unbounded effect on $\hat{\sigma}_V$. However, a minimax theory for robust scale estimation, similar to the theory of location estimation presented in Section 2.1.3, exist in the literature [51]. In this work, the scale of the noise is addressed as a nuisance parameter whose function is to adjust the clipping points of the parametric score functions in Figure 2.1 for estimating $\boldsymbol{\theta}$. Since wrong scaling of the score functions usually leads to either a decrease of efficiency or loss of robustness, it is important to use a robust scale estimator. Furthermore, it is desirable to use a simple plug-in method for scale estimation and put more effort in the location estimators [51]. The median or mean absolute deviation (mad) seems convenient for the above mentioned requirements and is defined for a sample $\mathbf{v}$ of M points as

$$\mathsf{mad}(\mathbf{v}) = \mathsf{E}\{|\mathbf{v} - \mathsf{median}(\mathbf{v})|\}, \quad (2.18)$$

where the expectation operator $\mathsf{E}\{\cdot\}$ can be replaced by the sample median or the sample mean. In the first case (2.18) is called median absolute deviation and in the second case it is called mean absolute deviation.

Note that the breakdown point of 50% of the median is inherited by the median absolute deviation [49]. To ensure consistency (2.10) under the Gaussian assumption a normalization constant is necessary yielding the robust scale estimator

$$\hat{\sigma}_V = 1.4826 \cdot \mathsf{mad}(\mathbf{v}). \quad (2.19)$$

However, the drawback of the median absolute deviation is that it yields low efficiency for small samples. Note that in practice, the noise sample $\mathbf{v}$ is not available and needs to be estimated.

2.1.4 Adaptive Estimation

An alternative to classical robust estimation, explained in Section 2.1.3, is to use adaptive techniques that incorporate more information of the observations into the estimator. The idea behind adaptive estimation is that instead of using a fixed bounded score function as described in Section 2.1.3, the score function of the true model is either approximated by a linear combination of basis functions, as explained in Section 2.1.4.1, or estimated non-parametrically. For the latter approach kernel density estimation (KDE), detailed in Section 2.1.4.2, is used throughout this thesis. In both cases the approximation of the score function is based on the residuals, i.e.,

$$\hat{v}_m = y_m - h_m(\hat{\boldsymbol{\theta}}), \qquad m = 1, 2, \ldots, M, \tag{2.20}$$

where $\hat{\boldsymbol{\theta}}$ is an initial (consistent) estimate, e.g., least-squares for symmetric distributions. The residuals are used to incorporate an estimate of the model $f_V(v)$ into the estimator. Then, the estimate of the model yields an improved estimate of the parameter $\boldsymbol{\theta}$. Model and parameter estimation is processed in an iterative loop until convergence is achieved.

The advantage of these approaches is that the estimators are able to adapt to the underlying noise environments (e.g., LOS or NLOS) and consequently achieve higher efficiency than conventional robust estimators given that the data size is large enough. The drawbacks are that higher computational power and in many (but not all) cases rather larger sample sizes are needed for the estimators to perform well. In this section, noise-adaptive estimation techniques assuming a symmetric $\mathcal{H}(v)$ are explained. Algorithms for these techniques and an extension to an asymmetric $\mathcal{H}(v)$ are presented in Chapter 3 and Chapter 4.

2.1.4.1 Adaptive Parametric Estimation

In [6, 12] the residuals $\hat{\mathbf{v}} = [\hat{v}_1, \hat{v}_2, \ldots, \hat{v}_M]^\mathsf{T}$ in (2.20) that contain information of the noise pdf are incorporated into the score function to improve estimation accuracy of the parameter $\boldsymbol{\theta}$. For this purpose, the score function is modeled as a linear combination of basis functions

$$\psi(v) = \sum_{j=1}^{J} u_j \phi_j(v) = \mathbf{u}^\mathsf{T} \boldsymbol{\phi}(v), \tag{2.21}$$

where u_j are the coefficients contained in vector $\mathbf{u} = [u_1, \ldots, u_J]^\mathsf{T}$ of the basis functions $\phi_j(v)$ contained in vector $\boldsymbol{\phi}(v) = [\phi_1(v), \ldots, \phi_J(v)]^\mathsf{T}$ chosen to be close to the expected model. Estimates of the weights **u** can be obtained by using the residuals and minimizing the mean square error (MSE) between the true score function and the estimated score function, i.e.,

$$\hat{\mathbf{u}} = \arg\min_{\mathbf{u}} \mathsf{E}\{(\psi(v) - \varphi(v))^2\}. \tag{2.22}$$

Under the assumption that

$$\lim_{v \to \pm\infty} \phi_j(v) f_V(v) = 0, \quad j = 1, 2, \ldots, J \tag{2.23}$$

one deduces that [6]

$$\hat{\mathbf{u}} = \mathsf{E}\{\boldsymbol{\phi}(v)\boldsymbol{\phi}^\mathsf{T}(v)\}^{-1}\mathsf{E}\{\boldsymbol{\phi}'(v)\}, \tag{2.24}$$

where $\boldsymbol{\phi}'(v)$ denotes the first order derivative of $\boldsymbol{\phi}(v)$ with respect to v. In practice the expectation operator $\mathsf{E}\{\cdot\}$ is replaced by the sample mean and v in (2.24) is replaced by the residuals $\hat{v}$ yielding the coefficients of the basis function which allows us to better approximate the score function $\psi(v)$. This function can then used in (2.12) to estimate the position of the UE. Nuisance parameters such as the scale of the noise residuals have to be estimated and plugged into the basis functions. However, even though the approach provides more flexibility than conventional robust schemes in Section 2.1.3, there are several limitations:

First, the performance of the estimator is highly sensitive to the choice of family of selected distributions. Second, even within one family of densities, it is not clear how many bases to include in the set (i.e. choice of J). If the number of bases is too large a higher variance of the estimated score function and the parameter estimate is expected. However, if the number of bases is too small to approximate the true score function it is likely that a bias occurs due to mis-modeling. Thus, the asymptotic variance of the estimator given in [12] is highly sensitive to the choice of the bases and weights. A model selection approach for finding a parsimonious set of bases is proposed in [42] in the context of MUD in impulsive noise, where it is found that one basis ($J = 1$) best trades off efficiency versus complexity. Note that for one basis this approach comes down to conventional robust estimation explained in the previous section. Third, although the basis function approach involves many tuning parameters, large deviations from the presumed model cannot be modeled. Note that in some cases, the basis function approach can increase estimation accuracy with respect to conventional robust techniques in Section 2.1.3, but it is conceptually and computationally heavy and not treated further throughout this thesis.

2.1.4.2 Semi-Parametric Estimation

In practice it is desirable for the estimator to have more degrees of freedom in order to adapt to the underlying situation, meaning to an LOS or NLOS scenario. An alternative to the previous subsection, where the score function is modeled parametrically, is to approximate the true score function in a non-parametric way. This approach usually provides more accurate approximations of the true score function which is expected to improve estimation accuracy of the parameter estimates. Furthermore, it is shown here that the number of tuning parameters with respect to the parametric approach can be reduced significantly. The proposed method combines non-parametric estimation of the noise pdf with an M-estimation procedure. Such approaches are termed semi-parametric, since they combine non-parametric estimation of an infinite-dimensional nuisance parameter (the noise pdf) with estimation of a finite-dimensional parameter of interest $\boldsymbol{\theta}$ [8]. Note that under certain conditions the semi-parametric estimator achieves the CRLB asymptotically for any noise pdf [8]. This is not true for small samples since a bias in the pdf estimate can result in position estimates with a higher RMSE.

The residuals (2.20) are used to compute non-parametric estimates (e.g. using KDE) $\hat{f}_V(v)$ and $\hat{f}'_V(v)$ of respectively $f_V(v)$ and its derivative $f'_V(v)$, yielding the score function estimate

$$\hat{\varphi}(v) = -\frac{\hat{f}'_V(v)}{\hat{f}_V(v)}. \qquad (2.25)$$

The estimated score function can then be used to estimate the parameter of interest $\boldsymbol{\theta}$ based on the ML principle, as in (2.5). Several KDE methods to estimate $f_V(v)$ exist in the literature [7, 27, 93]. The conventional KDE for a sample of points $(w_m)_{m=1}^M$ is defined as

$$\hat{f}_W(w) = \frac{1}{M\delta} \sum_{m=1}^{M} \mathcal{K}\left(\frac{w - w_m}{\delta}\right), \qquad (2.26)$$

for some chosen kernel density $\mathcal{K}(\cdot)$ (usually the standard normal density $\mathcal{N}(w; 0, 1)$) and a scale parameter, δ, also called bandwidth or smoothing parameter, that must be selected according to the available data. When $f_V(v)$ is assumed symmetric, we can implement the this constraint by symmetrizing the noise samples which doubles the sample size for KDE. This also ensures consistency since the estimated score function is anti-symmetrized (i.e. $\hat{\varphi}(-x) = -\hat{\varphi}(x)$). Applying (2.26) to the problem at hand using a single scale parameter δ at all points fails, since outliers produce spurious peaks at the tails of the distribution. This increases the number of solutions to the estimation equation (2.5) which can result in convergence problems when applying an optimization algorithm. This effect is illustrated in Figure 2.2 where KDE in (2.26) is used to estimate the pdf of a noise sample $\mathbf{v} = [v_1, v_2, \ldots, v_M]^\mathsf{T}$ that is distributed according to (2.2) where $\varepsilon = 0.4$, $\mathcal{H}(v) = \mathcal{N}(v; 0, 100\sigma_G^2)$ and σ_G^2 is chosen such that σ_V^2 equals one.

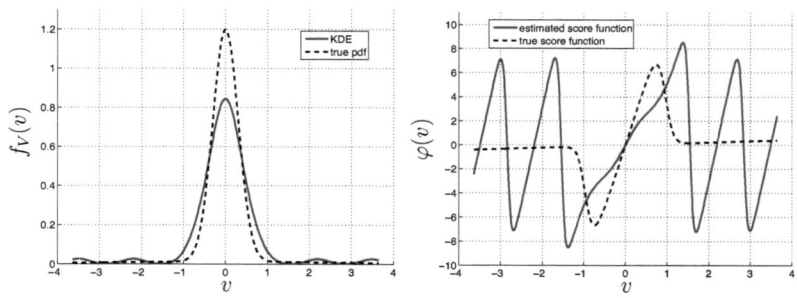

(a) $f_V(v)$ (dashed line) and its estimate (plain line) using KDE

(b) $\varphi(v)$ (dashed line) and its estimate (plain line) using KDE

Figure 2.2: $\hat{f}_V(v)$ is obtained using KDE according to Equation (2.26) from a noise sample of $M = 10$ data points, distributed according to (2.2) where $\varepsilon = 0.4$, $\mathcal{H}(v) = \mathcal{N}(v; 0, 100\sigma_G^2)$ and σ_G^2 is chosen such that σ_V^2 equals one.

It can be observed that bumps due to outliers occur at the tails of the distribution, resulting in oscillations of the estimated score function which lead to multiple solutions in the parameter estimation problem in (2.12). This problem is partly overcome in [113], where an estimator based on adaptive kernel density estimation (AKDE) is considered. In this approach KDE with local bandwidth δ_i is performed instead of only choosing a global one. Correction of local bandwidths for heavy tails in $\hat{f}_V(v)$ is performed using an additional order statistic of the sample. Moreover, $\hat{f}_V(v)$ is constrained to unimodality, which prevents ambiguities in solving (2.5) that may result from a multimodal $\hat{f}_V(v)$. This constraint is implemented by increasing the global bandwidth δ by a factor until a unimodal density is achieved [93]. Local bandwidth selection is thus a complex procedure that depends on the selection of several tuning parameters. Other approaches for local bandwidth selection may be used, see e.g. [93], but none of them can be defined in an optimal sense [26]. An estimator that uses a different local bandwidth selection rule would also require complex tail correction, to be applied in (2.25) to solve (2.5).

A conceptually and numerically simpler approach based on transformation density estimation [104], developed during this work, [44, 46] achieves performance similar to

2.1 Problem Statement for Geolocation

the estimator in [113] in symmetric noise environments. This approach is followed here. Since the residuals are likely to come from a heavy-tailed distribution $f_V(v)$, they are transformed by a nonlinear, parametric function $g(v,\zeta)$ such that they are assembled closer together after transformation. Symmetrizing the transformed residuals allows for conventional KDE (2.26) in the transformed domain (denoted as W-domain in the sequel). It also doubles the sample size which improves accuracy of the pdf estimate. An estimate of the original pdf is then obtained by back transformation, using

$$\hat{f}_V(v) = \hat{f}_W(g(v,\zeta)) \left| \frac{d\, g(v,\zeta)}{d\,v} \right|, \quad (2.27)$$

where $w = t(v,\zeta)$ and $f_W(w)$ is the density of the transformed data. This approach termed transformation kernel density estimation (TKDE) smoothes the pdf estimate by avoiding local maxima and oscillations at the tails of the distribution. To illustrate this effect TKDE estimation is performed with the same sample previously used and results for TKDE estimate and the corresponding score function are shown in Figure 2.3.

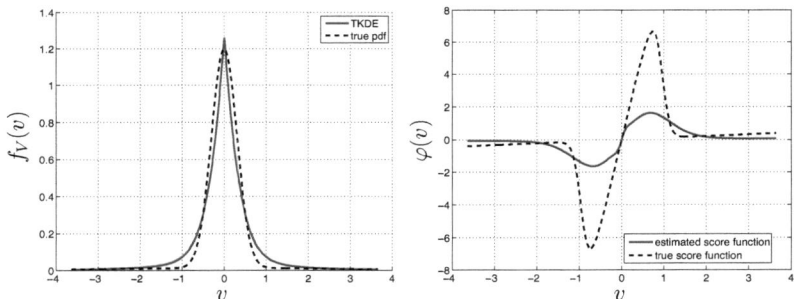

(a) $f_V(v)$ (dashed line) and its estimate (plain line) using TKDE

(b) $\varphi(v)$ (dashed line) and its estimate (plain line) using TKDE

Figure 2.3: $\hat{f}_V(v)$ is obtained using KDE according to Equation (2.26) from a noise sample of $M = 10$ data points, distributed according to (2.2) where $\varepsilon = 0.4$, $\mathcal{H}(v) = \mathcal{N}(v; 0, 100\sigma_G^2)$ and σ_G^2 is chosen such that σ_V^2 equals one.

We observe that the approach based on TKDE allows to obtain a smooth pdf estimate with one maximum when using an appropriate transformation function that gathers the samples closer together. This results in an estimated score function with a unique zero close to the true one and hence stabilizes parameter estimation in (2.12). In [104] the authors suggest that TKDE can be applied to change the tail behavior and peakedness of a sample and estimate the pdf with a complexity that is equivalent to that of conventional KDE using a global bandwidth. Two different transformation functions and parameter selection schemes are proposed below.

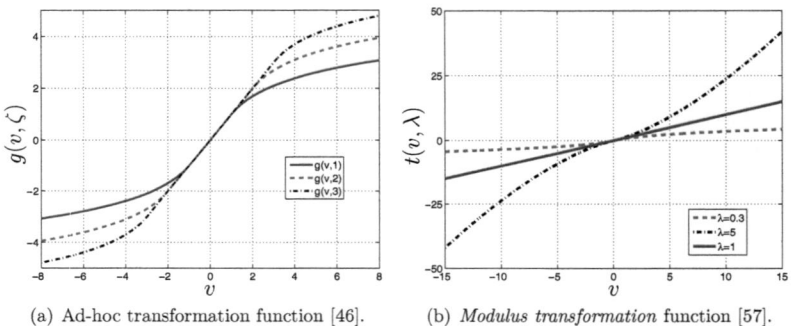

(a) Ad-hoc transformation function [46]. (b) *Modulus transformation* function [57].

Figure 2.4: Transformation functions used for correcting kurtosis of a symmetric sample.

Ad-hoc Modification of Residuals

In [46] an approach related to the idea of TKDE [104] for estimating $f_V(v)$ and its derivative is proposed which only requires KDE using a global bandwidth (2.26). In this construction, the original residuals $\hat{\mathbf{v}}$ are "tailored" by a parametric function into $\mathbf{w} = g(\hat{\mathbf{v}}, \zeta)$, where a cut-off parameter ζ is selected based on robust outlier detection [86]. The transformation function is linear with slope one in a certain region around zero and logarithmic beyond this region, thus transforming large outliers closer to the core. This simplifies using KDE with a global bandwidth in the W-domain and $\hat{f}_V(v)$ and its derivative are obtained via back-transformation using (2.27). The function $g(v, \zeta)$ is [46]

$$w = g(v, \zeta) = \begin{cases} v, & |v| \leq \zeta \\ \text{sign}(v) \cdot [\log(|v| - (\zeta - 1)) + \zeta], & |v| > \zeta, \end{cases} \quad (2.28)$$

and depicted in Figure 2.4(a). The function $g(v, \zeta)$ is continuous and continuously differentiable up to the first order with respect to v and ζ. Note that residuals smaller than the threshold ζ remain untransformed and can be easily estimated using a global bandwidth. Note also that ζ is selected according to an ad-hoc method [46] based on outlier detection [86] and order statistics. If a unimodal density is expected the global bandwidth is increased until a unimodal density estimate is achieved.

The function $g(v, \zeta)$ seems appropriate for data that are slightly contaminated. However, when large outliers occur, the transformation does not provide enough flexibility to transform these outliers close enough to the core. Furthermore, the iteration scheme of the unimodal constraint can lead to a large global bandwidth that increases the bias of the pdf estimate. A smaller global bandwidth in the transformed domain is desirable to obtain an estimate of the pdf with a small bias [93]. Improvements can be achieved by taking another transformation function.

Transformation-Based Semi-parametric Estimation

In [44] a more compact and conceptually simpler method for estimating the score function in (2.25) is proposed based on TKDE [104] and an estimate of $f_V(v)$ is obtained via back-transformation (2.27). Unlike [104] where, among others, the power family of transformations is considered, here we focus on the *modulus transformation* which is suitable to deal with symmetric and heavy-tailed data [57]. Later in Chapter 3, the approach is extended to asymmetric data and applied to the geolocation problem based on TOA measurements. The *modulus transformation* [57], a monotonic and point-symmetric function, defined as

$$w = t(v, \lambda) = \begin{cases} \text{sign}(v)\frac{(|v|+1)^\lambda - 1}{\lambda}, & \lambda \neq 0, \\ \text{sign}(v)\log(|v|+1), & \lambda = 0, \end{cases} \quad (2.29)$$

tends to *almost* normalize the data [16, 57]. This function is illustrated in Figure 2.4(b). For $\lambda > 1$, $t(v, \lambda)$ is convex when $v > 0$ and concave when $v < 0$, i.e., the transformed sample will be expanded. For $\lambda < 1$, it is concave when $v > 0$ and convex when $v < 0$, i.e., the transformed sample will be more concentrated around zero. Note that $t(v, \lambda)$ is linear for $\lambda = 1$ and continuously differentiable with respect to v and λ.

The transformation (2.29) is applied to the residuals, KDE is performed in the W-domain to obtain $\hat{f}_W(w)$ and $\hat{f}'_W(w)$. Then, $\hat{f}_V(v)$ and $\hat{f}'_V(v)$ are obtained via back-transformation and used to compute (2.25) for estimating $\boldsymbol{\theta}$. The level of nonlinearity of the derived $\hat{\varphi}$ using (2.29) adapts to the data automatically via the selection of λ. In contrast to [46] no cut-off point is to be set and the selection of λ is done by an MLE based on the assumption that the data after transformation is Gaussian [57]. Replacing $f_W(w)$ in (2.27) by the Gaussian pdf with mean μ_W and variance σ_W^2 and $g(v, \zeta)$ by $t(v, \lambda)$ yields

$$f_V(v) = \frac{1}{\sqrt{2\pi}\sigma_W} \exp\left\{-\frac{(t(v, \lambda) - \mu_W)^2}{2\sigma_W^2}\right\} (|v|+1)^{\lambda - 1}. \quad (2.30)$$

For iid observations, the log-likelihood function is constructed by replacing μ_W and σ_W^2 by their sample mean and variance estimates of the transformed data $\mathbf{w}$. Discarding the terms that do not contain λ, one obtains

$$\hat{\lambda} = \arg\max_\lambda \left\{-\frac{M}{2}\log \hat{\sigma}_W^2(\lambda) + (\lambda - 1)\sum_{m=1}^M \log(|v_m|+1)\right\}. \quad (2.31)$$

In order to transform the data closer together, we set $\lambda < 1$ to constrain $t(v, \lambda)$ to be concave for $v > 0$ and convex for $v < 0$, thus suppressing the need for local bandwidth selection. Several other (robust and non-parametric) estimators for λ exist in the literature [16, 50], and may be adapted to the *modulus transformation*. However, the impact of the choice of λ on the parameter estimator does not seem critical for a symmetric heavy-tailed noise distribution, as long as λ is constraint to be smaller than one [44]. MLE for λ is used for its simplicity compared to other methods. Note that our goal is not to estimate $f_V(v)$ but the parameter $\boldsymbol{\theta}$. Note also that the Gaussian assumption is only made for the step of the selection of λ. By [16, 57] one can assume that the data transformed by the *modulus transformation* are *almost* Gaussian, which is done for selecting λ but avoided for determining $f_W(w)$ in (2.27) in order to remain adaptive. Symmetrization of the residuals is applied in the transformed domain and the pdf in the original domain inherits the smoothness property of the pdf in the transformed domain via $t(v, \lambda)$. It was found that the plug-in rule for a global bandwidth $\hat{\delta} = 1.06 \hat{\sigma}_W M^{-1/5}$ [93] is suitable here for KDE in the W-domain. Unlike the scheme in [46], a unimodal density is often achieved for a broader choice of bandwidth selection rules. This reduces bias of the pdf estimate and computational load compared to [46]. Depending on the noise environment, the constraint of unimodality can be left out in the proposed algorithm.

Note that, besides the pdf estimate, there are only two nuisance parameters to estimate (the noise scale σ_W and λ) for determining the parameter of interest $\boldsymbol{\theta}$. The asymptotic variance of the semi-parametric estimator has not yet been established. Resampling techniques, e.g., [114], can be used to obtain accurate estimates of its asymptotic covariance.

2.2 Problem Statement for Tracking

In many applications such as yellow pages services or fleet management, the UE is moving over time and the aim is to determine the position and velocity of the UE based on signal parameters at different time steps, i.e., updating the previous estimates as soon as new measurements are available. Using recursive or sequential algorithms that take into account estimates from the previous time step as well as the actual incoming measurements are appealing for two reasons: First, sequential processing saves computational power with respect to batch-processing since only the new measurements, as opposed to all measurements, are used for updating the previous estimate. Second, position errors can be decreased thanks to time averaging effects and a gain in performance can be achieved with respect to the static case discussed in Section 2.1.

Unlike for batch-processing, described in Section 2.1, where the parameters of interest are deterministic, here, in a recursive Bayesian estimation framework, the quantities of interest are random variables (rvs).

2.2.1 Signal Model for a Moving User Equipment

For notational convenience the rv in the subscript of the density $f_V(v)$ is dropped, i.e., $f_V(v) = f(v)$. A capital letter in the argument of $f(\cdot)$ denotes an observed time sequence, i.e., $\mathbf{Y}^k = \{\mathbf{y}(i),\ i = 1,\ldots,k\}$.

2.2.1.1 Nonlinear System Model

Consider that the UE is moving on a 2D plane surrounded by M FTs. The parameters of interest are the position and velocity of the UE at each time step k contained in the state vector $\mathbf{x}(k) = [x(k)\ y(k)\ \dot{x}(k)\ \dot{y}(k)]^\mathsf{T}$, where $\dot{x}(k)$ and $\dot{y}(k)$ denote the derivatives of $x(k)$ and $y(k)$ with respect to time, corresponding to the velocity in x- and y-direction. The following discrete-time stochastic model describes the movement of the UE, i.e.,

$$\mathbf{x}(k) = \mathbf{a}(\mathbf{x}(k-1)) + \boldsymbol{\omega}(k-1), \quad k = 1, 2, \ldots, K, \tag{2.32}$$

where $\mathbf{a}(\cdot)$ is a known linear or nonlinear function describing the movement of the UE from time step $k-1$ to k, and vector $\boldsymbol{\omega}(k-1)$ is white Gaussian process noise with covariance matrix $\mathbf{Q}(k)$, $k = 1,\ldots,K$, describing the uncertainty one has about the motion model. Thus, $\mathbf{x}(k)$ is a first-order Markov process meaning that $\mathbf{x}(k)$ only depends on $\mathbf{x}(k-1)$ and $\boldsymbol{\omega}(k-1)$. Consequently, the condition $f(\mathbf{x}(k)|\mathbf{x}(k-1),\mathbf{x}(k-2),\ldots,\mathbf{x}(0)) = f(\mathbf{x}(k)|\mathbf{x}(k-1))$ holds, where $f(\mathbf{x}(k)|\mathbf{x}(k-1))$ is the pdf of the process noise. Various motion models with different $\mathbf{a}(\cdot)$ to describe different dynamics exist in the literature [85].

Since the UE is moving, the nonlinear model from (2.1) becomes time-dependent, i.e.,

$$\mathbf{y}(k) = \mathbf{h}(\mathbf{x}(k)) + \mathbf{v}(k), \quad k = 1, 2, \ldots, K. \tag{2.33}$$

where $\mathbf{h}(\cdot)$ describes the nonlinear relationship between the position of the UE and the FTs as explained in Section 2.1.1. The measurement noise sequence $\mathbf{v}(k) = [v_1(k), v_2(k), \ldots, v_M(k)]^\mathsf{T}$ is assumed white over time and FTs, and each element $v_m(k)$ is distributed according to (2.2). The measurement covariance matrix $\mathbf{R}(k) = \mathsf{E}\{(\mathbf{v}(k) - \mathsf{E}\{\mathbf{v}(k)\})(\mathbf{v}(k) - \mathsf{E}\{\mathbf{v}(k)\})^\mathsf{T}\}$ can be time-variant. For the signal parameters TOA, AOA and RSS and a fixed time step k, given that $\mathbf{v}(k)$ is iid among FTs, $\mathbf{R}(k) = \sigma_V^2 \mathbf{I}_M = (\sigma_G^2 + \varepsilon \sigma_\eta^2)\mathbf{I}_M$, where $\mathbf{I}_M$ is the $M \times M$ identity matrix. Note that $\mathbf{R}(k)$ is not a diagonal matrix for TDOA positioning. This is because a reference FT is chosen and the time differences to all FTs are calculated leading to correlations among the different pairs. Note that $\mathbf{v}(k)$ and $\boldsymbol{\omega}(k)$ are mutually independent for all k.

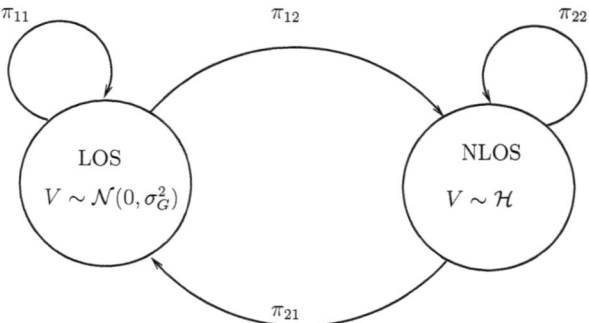

Figure 2.5: Markov chain for modeling LOS/NLOS occurrences.

2.2.1.2 Jump-Markov Nonlinear Model

Unless the UE is maneuvering quickly, which is beyond the scope of this work, the whiteness assumption for the process noise in (2.32) is reasonable. In contrast, the assumption for a white process $\mathbf{v}(k)$ (2.33) may not hold in reality since the wireless channel for moving objects highly depends on the environments. Thus, the perturbations due to the NLOS effect can undergo switching over time, meaning at time step $k-1$ an FT can be in LOS and at time step k it changes to NLOS or vice versa. Imagine that a UE is driving alongside an FT shadowed by a high-rise building. During this time, consecutive measurements are in NLOS condition. When the building is passed, the condition of the channel abruptly changes to LOS. The opposite, i.e., consecutive measurements are in LOS condition, is true for a UE traveling on a flat plane where no obstacles prevent the signals to arrive at the FT via the direct path. In this case a sudden change of the channel to NLOS occurs when the UE passes an obstacle.

To model these time dependencies and sudden changes, for each FT we use a first-order time-homogeneous Markov chain [25], depicted in Figure 2.5, consisting of $r = 2$ states. Note that a process $\mathcal{M}(k)$ is called Markov chain (MC) when the future of the process only depends on the present and not on the past. In the following $\mathcal{M}(k)$ is called the mode variable, assumed to be among the r possible modes, $\mathcal{M}(k) \in \{\mathcal{M}_j\}_{j=1}^r$, where $\mathcal{M}_1$ is assigned to the event "LOS" and $\mathcal{M}_2$ is assigned to the event "NLOS". The transition probabilities $\pi_{ij} \geq 0$ denote the conditional probability for changing to state $\mathcal{M}_j$ at time step k given that state $\mathcal{M}_i$ is in effect at $k-1$,

$$\pi_{ij} = \Pr\{\mathcal{M}(k) = \mathcal{M}_j | \mathcal{M}(k-1) = \mathcal{M}_i\}, \quad \forall\, i,j = 1,\ldots,r, \qquad (2.34)$$

2.2 Problem Statement for Tracking

and the transition probability matrix (TPM) of the MC depicted in Figure 2.5 is

$$\mathbf{\Pi} = \begin{bmatrix} \pi_{11} & \pi_{12} \\ \pi_{21} & \pi_{22} \end{bmatrix}, \tag{2.35}$$

where $\sum_{j=1}^{2} \pi_{ij} = 1$, $i = 1, 2$. Given any initial probabilities, meaning the probabilities that the MC at time step $k = 1$ is in LOS, i.e., $\Pr\{\mathcal{M}(1) = \mathcal{M}_1\} = \pi_1$ and $\Pr\{\mathcal{M}(1) = \mathcal{M}_2\} = 1 - \pi_1$, where $0 \leq \pi_1 \leq 1$, then the MC converges to a stationary distribution when k increases [25],

$$\lim_{k \to \infty} \Pr\{\mathcal{M}(k) = \mathcal{M}_1\} = \frac{\pi_{21}}{\pi_{12} + \pi_{21}} = 1 - \varepsilon, \tag{2.36}$$

$$\lim_{k \to \infty} \Pr\{\mathcal{M}(k) = \mathcal{M}_2\} = \frac{\pi_{12}}{\pi_{12} + \pi_{21}} = \varepsilon. \tag{2.37}$$

However, considering M FTs, there are $r = 2^M$ different noise constellations or modes, meaning the MC consists of 2^M different states. Assuming that the LOS/NLOS transitions among different FTs are independent we can calculate the TPM for the augmented MC (meaning the whole system with M FTs) using the Kronecker product ($\otimes$), i.e.,

$$\mathbf{T} = \mathbf{\Pi}_1 \otimes \mathbf{\Pi}_2 \otimes, \ldots, \otimes \mathbf{\Pi}_M, \tag{2.38}$$

where $\mathbf{\Pi}_m$ is the TPM (2.35) of the m-th FT and $\mathbf{T}$ with dimensions $r \times r$ is the TPM of the augmented MC. The elements of $\mathbf{T}$ are the transitions probabilities between the r different states. Since (2.1) is now time-dependent, the time-varying LOS/NLOS occurrences are modeled as a jump-nonlinear system [5] with discrete mode variable $\mathcal{M}(k) \in \{\mathcal{M}_j\}_{j=1}^{r=2^M}$ describing the mode the system is in at time step k,

$$\mathbf{y}(k) = \mathbf{h}(\mathbf{x}(k)) + \mathbf{v}(k, \mathcal{M}(k)) \quad k = 1, 2, \ldots, K, \tag{2.39}$$

Such systems are called hybrid because they consist of continuous components, such as the state vector $\mathbf{x}(k)$, and the discrete mode variable $\mathcal{M}(k)$. Note that, $\mathbf{v}(k, \mathcal{M}(k))$ and $\boldsymbol{\omega}(k)$ in (2.32) and (2.39) are assumed mutually independent. To model maneuver, the mode variable $\mathcal{M}(k)$ can also be incorporated into the process noise in (2.32). Here, we only assume $\mathbf{v}(k, \mathcal{M}(k))$ to be mode-dependent to model NLOS effects with $\mathbf{R}(k) = \mathsf{E}\{(\mathbf{v}(k, \mathcal{M}(k)) - \mathsf{E}\{\mathbf{v}(k, \mathcal{M}(k))\})(\mathbf{v}(k, \mathcal{M}(k)) - \mathsf{E}\{\mathbf{v}(k, \mathcal{M}(k))\})^\mathsf{T}\}$. Thus, the measurement covariance matrix is strongly time-dependent and changes between consecutive time steps due to LOS/NLOS switching according to the augmented MC in (2.38). Note that for TOA, AOA, RSS the diagonal elements of $\mathbf{R}(k)$ at a fixed time step are equal either to σ_G^2 or to $\sigma_G^2 + \sigma_\eta^2$ and do not depend on ε anymore. As in Section 2.2.1.1 the structure of $\mathbf{R}(k)$ is non-diagonal for TDOA. Note also that the process noise covariance $\mathbf{Q}(k)$ is assumed as in Section 2.2.1.1. The algorithms for tracking a UE, developed in Chapter 4 and Chapter 5, are tested under both, the nonlinear measurement model in (2.33) and the jump-nonlinear model in (2.39).

2.2.2 Nonlinear Filtering

The aim of nonlinear filtering is to recursively estimate the unknown pdf[3] $f(\mathbf{x}(k)|\mathbf{Y}^k)$ using the observations $\mathbf{y}(k)$, the state space equations (2.32) and (2.33) and the known pdfs of the noise sequences $\boldsymbol{\omega}(k)$ and $\mathbf{v}(k)$. Then, the optimal solution for the state vector, containing position and velocity of the UE, in the maximum a posteriori (MAP) or minimum mean-square error (MMSE) sense can be calculated. The former is based on maximizing the posterior pdf $f(\mathbf{x}(k)|\mathbf{Y}^k)$ whereas the latter is based on calculating the conditional expectation $\mathsf{E}\{\mathbf{x}(k)|\mathbf{Y}^k\}$. A necessary condition for the sequel is that $\mathbf{v}(k)$ and $\mathbf{w}(k)$ are mutually independent, white sequences. Further, assume that the initial state vector has known pdf $f(\mathbf{x}(k-1)|\mathbf{Y}^{k-1})$, denoted as prior (describes a-priori information of the state vector before any measurements are collected) and that it is independent of the noise sequences $\mathbf{v}(k)$ and $\boldsymbol{\omega}(k)$.

Since $\mathbf{x}(k)$ is Markov, $f(\mathbf{x}(k)|\mathbf{x}(k-1), \mathbf{Y}^{k-1}) = f(\mathbf{x}(k)|\mathbf{x}(k-1))$ which corresponds to the process noise pdf evaluated at $\mathbf{x}(k) - \mathbf{a}(\mathbf{x}(k-1))$. Then, the prediction density can be obtained by using the Chapman-Kolmogorov equation, i.e.,

$$f(\mathbf{x}(k)|\mathbf{Y}^{k-1}) = \int f(\mathbf{x}(k)|\mathbf{x}(k-1))f(\mathbf{x}(k-1)|\mathbf{Y}^{k-1})d\mathbf{x}(k-1). \qquad (2.40)$$

When a new measurement becomes available at time step k, the update step is performed. Using Bayes' Theorem we obtain an update of the prediction density

$$\begin{aligned}
f(\mathbf{x}(k)|\mathbf{Y}^k) &= f(\mathbf{x}(k)|\mathbf{y}(k), \mathbf{Y}^{k-1}) & (2.41)\\
&= \frac{f(\mathbf{y}(k)|\mathbf{x}(k), \mathbf{Y}^{k-1})f(\mathbf{x}(k)|\mathbf{Y}^{k-1})}{f(\mathbf{y}(k)|\mathbf{Y}^{k-1})} & (2.42)\\
&= \frac{f(\mathbf{y}(k)|\mathbf{x}(k))f(\mathbf{x}(k)|\mathbf{Y}^{k-1})}{\int f(\mathbf{y}(k)|\mathbf{x}(k))f(\mathbf{x}(k)|\mathbf{Y}^{k-1})d\mathbf{x}(k)}, & (2.43)
\end{aligned}$$

where $f(\mathbf{y}(k)|\mathbf{x}(k), \mathbf{Y}^{k-1}) = f(\mathbf{y}(k)|\mathbf{x}(k))$ denotes the measurement noise pdf evaluated at $\mathbf{y}(k) - \mathbf{h}(\mathbf{x}(k))$.

At each time step, one can compute the optimal MMSE estimator using the recursions above, i.e.,

$$\hat{\mathbf{x}}(k|k)^{\mathsf{MMSE}} = \mathsf{E}\{\mathbf{x}(k)|\mathbf{Y}^k\} = \int \mathbf{x}(k)f(\mathbf{x}(k)|\mathbf{Y}^k)d\mathbf{x}(k), \qquad (2.44)$$

that achieves the Posterior Cramér-Rao Lower Bound (PCRLB). Similar to the CRLB in Section 2.1.2, the PCRLB can be computed using the posterior distribution $f(\mathbf{x}(k)|\mathbf{Y}^k)$ [84] and can serve as a benchmark to compare the posterior covariance of any unbiased state estimator. If $\mathbf{a}(\cdot)$ and $\mathbf{h}(\cdot)$ are linear, and $\boldsymbol{\omega}(k)$ and $\mathbf{v}(k)$ are white zero-mean Gaussian and mutually independent, a closed-form recursion of Equations

---

[3]In contrast, in Section 2.1.4.2 a non-parametric pdf estimate of $f(\mathbf{y}(k)|\boldsymbol{\theta})$ is obtained by considering the whole measurements at one particular time step.

2.2 Problem Statement for Tracking

(2.40)-(2.43) exist which is known as the Kalman filter (KF) [1]. In this case the MMSE and MAP estimators coincide. However, if one of the functions $\mathbf{h}(\cdot)$ or $\mathbf{a}(\cdot)$ is nonlinear or the densities of $\boldsymbol{\omega}(k)$ or $\mathbf{v}(k)$ are non-Gaussian, an analytical solution for $f(\mathbf{x}(k)|\mathbf{Y}^k)$ becomes intractable in most cases. The former is even true for wireless positioning in LOS environments since the relationship $\mathbf{h}(\cdot)$ between the FTs, the UE and the observations $\mathbf{y}(k)$ is nonlinear. Furthermore, if NLOS propagation occurs, the density of $\mathbf{v}(k)$ is non-Gaussian. To obtain an optimal solution in these cases requires knowledge of the distribution of $\mathbf{v}(k)$ and we can use numerical integration for solving (2.40) and (2.43). This implies keeping track of the different pdfs which can be a computational burden for higher order problems. Therefore, optimal solutions are computationally expensive and difficult to implement and suboptimal approaches are preferred in many problems [1, 84].

Suboptimal filters include the extended Kalman filter (EKF) and unscented Kalman filter (UKF), among others [84] that achieve good performance in LOS environments [53]. Both filters are based on the same assumptions as the KF mentioned above except that the functions $\mathbf{a}(\cdot)$ or $\mathbf{h}(\cdot)$ are nonlinear. Since the Gaussian pdf is completely described by the first and second moments, the EKF and UKF only require to keep track of the state vector and its covariance instead of the entire pdf. For the EKF, the nonlinear functions $\mathbf{a}(\cdot)$ and $\mathbf{h}(\cdot)$ are approximated by a first-order Taylor series. Similar to the recursions in (2.40)-(2.43), the EKF consists of prediction and update steps and one cycle of the algorithm is illustrated in Figure 2.6.

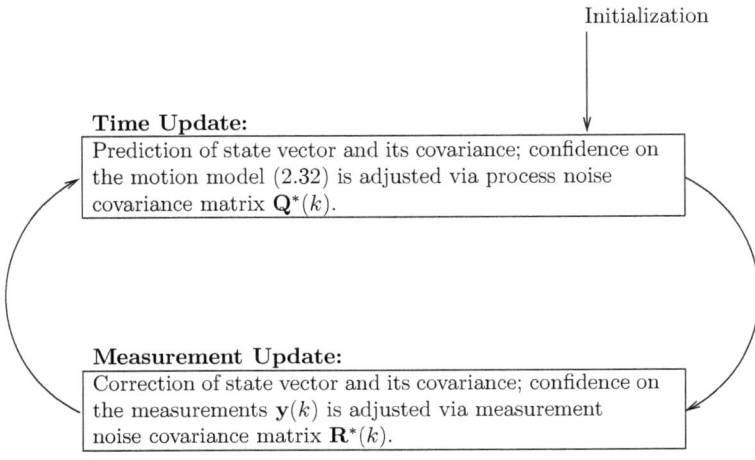

Figure 2.6: One cycle of the KF, EKF and UKF.

Based on an initial state vector (drawn from a known Gaussian distribution $f(\mathbf{x}(0)|\mathbf{Y}^0)$), the EKF predicts the state vector and the associated covariance according to the prediction equation (2.40) (Time Update). This equation comes down to a closed form under the assumption mentioned above and when $\mathbf{a}(\cdot)$ is approximated by a first-order Taylor series. Because of the linear relationship between the previous and the predicted state vector, the latter remains Gaussian. Then, when new measurements come in (Measurement Update), the predicted state vector and its covariance are corrected to follow the measurements. Again, (2.43) comes down to a closed form and the update of the state vector remains Gaussian given that the measurement noise $\mathbf{v}(k)$ is Gaussian- which is true for LOS.

The sensitivity of the EKF towards changes in the motion or measurement model is adjusted by the covariance matrices $\mathbf{Q}^*(k)$ and $\mathbf{R}^*(k)$ that need to be set since the true ones, $\mathbf{Q}(k)$ and $\mathbf{R}(k)$, given in Section 2.2.1, are unknown in practice. Since any deterministic mathematical motion model $\mathbf{a}(\cdot)$ does not capture the entire reality, it is necessary to adjust the uncertainty one has in the motion model by adapting the presumed covariance $\mathbf{Q}^*(k)$ of the process noise $\boldsymbol{\omega}(k)$. Larger covariance matrices $\mathbf{Q}^*(k)$ lead to higher uncertainty meaning less confidence in the model $\mathbf{a}(\cdot)$ whereas smaller covariance matrices $\mathbf{Q}^*(k)$ make $\mathbf{a}(\cdot)$ more trustable. The same is true for the measurement covariance $\mathbf{R}^*(k)$:

If it is chosen small enough, the EKF will overweigh the measurements with respect to the model leading to fast convergence. However, large outliers due to NLOS can seriously effect state estimation since they are considered as valid measurements. In contrast, if $\mathbf{R}^*(k)$ is chosen too large, the EKF rather trust the model $\mathbf{a}(\cdot)$ and only slightly learns from the measurements which leads to slow convergence speed or at worst to divergence. Hence, the choice of $\mathbf{Q}^*(k)$ and $\mathbf{R}^*(k)$ is always a trade-off between convergence speed and robustness towards deviations from the model assumptions.

More details on the EKF, which serves for comparison purposes in this thesis, is given in Chapter 4. If the functions $\mathbf{a}(\cdot)$, $\mathbf{h}(\cdot)$ are highly nonlinear, the EKF loses positioning accuracy due to the linearization errors. Improvements can be achieved using the UKF [58, 103]. This filter is based on transforming deterministically chosen sample points through the nonlinear functions and thus better approximates the true mean and covariance after transformation resulting in lower state estimation errors when $\mathbf{a}(\cdot)$ and $\mathbf{h}(\cdot)$ are highly nonlinear.

However, the main problem in this work is non-Gaussian measurement noise due to NLOS propagation. NLOS outliers affect the EKF and the UKF since both filters process the measurements linearly resulting in large state estimation errors [53, 55].

An alternative to EKF and UKF for non-Gaussian measurement noise are particle filters (PFs) which are more computational demanding [84]. A certain process and measurement noise model is assumed and the integrals in (2.40), (2.43) and (2.44) are solved using Monte Carlo methods for numerical integration. In situations, in which the presumed model fits reality, PFs can gain significantly in performance with respect to EKF and UKF.

However, the three suboptimal techniques mentioned above share the same limitations that they are not robust against large deviations from the assumed model. This is the Gaussian assumption for the EKF and UKF and any parametric model assumption for the PF. Thus, robust state estimation techniques are required to cope with uncertainties in the measurement model. These techniques are discussed in Section 2.2.4 and further developed in Chapter 4.

2.2.3 Hybrid Nonlinear Filtering

In the previous section we assumed that the noise sequences $\boldsymbol{\omega}(k)$ and $\mathbf{v}(k)$ are white and $f(\mathbf{x}(k)|\mathbf{Y}^k)$ is estimated recursively for determining the state vector $\mathbf{x}(k)$. Since the whiteness assumption (at least for $\mathbf{v}(k)$) is too stringent for practical applications, the jump-nonlinear model (2.39) is presented in Section 2.2.1.2 where LOS/NLOS occurrences for each FT are modeled by an MC, depicted in Figure 2.5. Based on this model, we want to recursively estimate the unknown pdf $f(\mathbf{x}(k)|\mathbf{Y}^k)$ using the observations $\mathbf{y}(k)$ and the state equation (2.32). Then, $f(\mathbf{x}(k)|\mathbf{Y}^k)$ can be used to estimate the state vector which yields the optimal solution in the MMSE or MAP sense, as explained in Section 2.2.2.

Recall the hybrid estimation problem with state equation (2.32) and measurement equation (2.39) where $\mathcal{M}(k)$ is a discrete-valued mode variable, modeling LOS/NLOS occurrences, and $\mathbf{x}(k)$ is the continuous-valued state vector. Mode jumps among the different models $\mathcal{M}_1, \mathcal{M}_2, \ldots, \mathcal{M}_r$, $r = 2^M$ occur randomly from time step $k-1$ to time step k. To model these changes, let $\mathcal{M}^{k,l}$ denote the l-th mode history at time step k. Then, the sequence of modes through time step k is

$$\mathcal{M}^{k,l} = \{\mathcal{M}_{i_{1,l}}, \ldots, \mathcal{M}_{i_{k,l}}\} \qquad l = 1, \ldots, r^k, \qquad (2.45)$$

where the index $i_{\kappa,l}$ denotes the model index at time κ from history l and $1 \leq i_{\kappa,l} \leq r$, $\kappa = 1, \ldots, k$ [5]. The number of mode histories increases with $(2^M)^k$. Consider $M = 1$ FT, at time step $k = 3$ there are $l = (2^1)^3$ different mode sequences given in Table 2.2. These sequences of model changes need to be incorporated into update and prediction densities from Section 2.2.2. Instead of considering (2.40) and (2.43), the switching at each time step is incorporated into the estimator, yielding the mode-conditioned prediction density for $l = 1, \ldots, r^k$, i.e.,

$$f(\mathbf{x}(k)|\mathcal{M}^{k-1,l}, \mathbf{Y}^{k-1}) = \int f(\mathbf{x}(k)|\mathbf{x}(k-1)) f(\mathbf{x}(k-1)|\mathcal{M}^{k-1,l}, \mathbf{Y}^{k-1}) d\mathbf{x}(k-1), \quad (2.46)$$

and the mode-conditioned posterior density

$$f(\mathbf{x}(k)|\mathcal{M}^{k,l},\mathbf{Y}^k) = \frac{f(\mathbf{y}(k)|\mathcal{M}^{k,l},\mathbf{x}(k))f(\mathbf{x}(k)|\mathcal{M}^{k-1,l},\mathbf{Y}^{k-1})}{\int f(\mathbf{y}(k)|\mathcal{M}^{k,l},\mathbf{x}(k))f(\mathbf{x}(k)|\mathcal{M}^{k-1,l},\mathbf{Y}^{k-1})d\mathbf{x}(k)}. \qquad (2.47)$$

Note that we assume that only the measurement equation is mode-dependent whereas the state equation (2.32) is not. The posterior density incorporating the priors with the transition probabilities from $\mathbf{T}$ then yields

$$f(\mathbf{x}(k)|\mathbf{Y}^k) = \sum_{l=1}^{r^k} f(\mathbf{x}(k)|\mathcal{M}^{k,l},\mathbf{Y}^k)\mathsf{Pr}\{\mathcal{M}^{k,l}|\mathbf{Y}^k\}, \qquad (2.48)$$

where $\mathsf{Pr}\{\mathcal{M}^{k,l}|\mathbf{Y}^k\}$ is calculated recursively [5]. Note that if the mode in effect at every time step k is known, the solution of the hybrid problem comes down to nonlinear filtering, explained in the previous Section 2.2.2.

l	$i_{1,l}$	$i_{2,l}$	$i_{3,l}$
1	1	1	1
2	1	1	2
3	1	2	1
4	1	2	2
5	2	1	1
6	2	1	2
7	2	2	1
8	2	2	2

Table 2.2: Mode sequences for one FT at time step $k = 3$.

An optimal solution for the hybrid system would require to model all sequences at any time step. Since the number of sequences is increasing with $(2^M)^k$, optimal solutions become intractable and suboptimal approximations are preferred [5, 71]. Unless a certain sequence of mode jumps is fixed, calculating the PCRLB becomes intractable for the same reasons [84].

Suboptimal solutions keep only a fixed number of sequences with highest probability and discard the rest. The probabilities of the remaining sequences are normalized such that they sum up to one. Then, depending on the approach a certain number of filters (e.g., KFs, EKFs or UKFs), matched to a certain model, operate in parallel. Suboptimal approaches for the hybrid estimation problem are the generalized pseudo-Bayesian (GPB) and interacting multiple model (IMM) algorithm [5]. The latter is most often used because it only requires r filters in parallel and trades off performance versus complexity in an appropriate manner [71]. Different IMM algorithms have been applied to the NLOS problem in [20, 33, 67]. However, in these papers, the authors assume entire knowledge of the NLOS error statistics which is not realistic for practical problems. Furthermore, even though the IMM algorithm well suits the jump-nonlinear model (2.39), the r filters, used within this algorithm (e.g. KFs, EKFs or UKFs) matched to the different models, remain highly sensitive to deviations from the assumptions. Thus, robustness of the IMM cannot be guaranteed without robustifying the filters used within this algorithm. Different approaches towards robust state estimators are presented in the next section, while a robustified IMM filter for tracking a UE is developed in Chapter 4 and published in [47].

2.2.4 State Estimation in the Presence of Outliers

2.2.4.1 General Problem

This section gives a brief overview of robust state estimators and related techniques in the presence of outliers. Here we distinguish two different kinds of outliers [69,87]: First, innovation outliers occur in the process noise $\boldsymbol{\omega}(k)$ and can be used, e.g., for modeling maneuvers. Second, measurement outliers occur in the measurement equations (2.33) or (2.39) and depend on the assumed model that describes perturbations due to the NLOS effect. Since we deal with NLOS error mitigation the former are not treated further throughout this work.

Non-Gaussian filtering for linear state and measurement equations is treated in [69,75]. In [69] the prediction density $f(\mathbf{y}(k)|\mathbf{Y}^{k-1})$ is assumed non-Gaussian and results from convolving the density of the measurement noise pdf $f(\mathbf{y}(k)|\mathbf{x}(k))$ with the pdf of the prediction error while in [75] a Student-t distribution is used to model the prediction density limiting the impact of large outliers. However, these state estimators strongly depend on the presumed model and thus provide only a limited degree of robustness. If the jump-nonlinear model from (2.39) is considered, the measurement noise pdf changes over time due to model changes between LOS and NLOS. Thus, applying a specific parametric model (which may be non-robust) for all time steps is not suitable.

Different approaches, based on clipping outliers in the KF innovations using a bounded score function, as explained in Section 2.1.3.1, are considered in [70,87] for linear state and measurement equations. In [87] it is proposed to choose the clipping parameters of the bounded score function in terms of the efficiency loss one is willing to pay under the nominal Gaussian assumption. The covariance matrix of the KF is computed offline to define the efficiency loss in the steady state. This is not possible for nonlinear problems since the covariance matrices depend on the measurements. Even though these approaches are well suited for linear problems, removing information from the measurements can result in convergence problems when nonlinear problems are considered.

In [99] an outlier-robust KF is suggested where the elements of the measurement covariance matrix $\mathbf{R}^*(k)$ are weighted according to the confidence one has on a particular measurement. Hence, large outliers increase the corresponding elements of $\mathbf{R}^*(k)$, thus reducing their effects on the state update. In contrast, the elements of $\mathbf{R}^*(k)$ corresponding well to the nominal data are reduced to emphasize their impact on state estimation.

A more detailed overview of different robust state estimation schemes can be found in [90].

As described in Section 2.1.3, an alternative to robust parametric or semi-parametric approaches is to perform outlier detection and estimate the parameters from the remaining data with concepts of missing observations. A KF based on missing observations is suggested in [94]. In [88] an outlier detection and state estimation scheme is proposed where the measurements are tested against outliers and only the remaining ones are used in the update step of the KF. The probabilistic data association (PDA) algorithm [62] is based on a similar concept designed to cope with measurement uncertainties. Large outliers are discarded and the remaining measurements are weighted with different probabilities in a KF framework. A modified probabilistic data association (MPDA) algorithm adapted to the UE tracking problem in NLOS environments is proposed in Chapter 5 and published in [48].

2.2.4.2 Robust Regression Kalman Filtering

The most popular approach to robust Kalman filtering is to rewrite the Kalman filter equations into a linear regression model and apply robust M-estimation techniques to solve for the state vector [9, 28, 87]. This approach is extended in [29] to nonlinear measurement equations and adopted in [80] to the problem of UE tracking. The clipping parameters of the bounded score function have to be estimated based on the measurement residuals in order to provide adaptivity for different noise situations. These approaches are robust over a class of distributions $\mathcal{H}(v)$ and serve as a benchmark for comparison purposes. Furthermore, the EKF equations in regression form are exploited in Chapter 4 to design a semi-parametric EKF [45] based on the concepts of semi-parametric estimation explained in Section 2.1.4.2. The proposed noise-adaptive EKF estimates the interference pdf at any time step non-parametrically, yielding smaller state estimation errors than its parametric counterparts.

A parametric alternative to the semi-parametric EKF is proposed in Chapter 4 and [47] where the IMM algorithm is adopted. An EKF together with a robust extended Kalman filter (REKF) based on robust regression are processed in parallel and the final state estimate is a combination of the individually computed state estimates yielding more robust solutions in NLOS and higher efficiency in LOS environments.

Chapter 3
Robust Geolocation

This chapter deals with the problem of finding the position of a stationary UE based on TOA measurements between the UE and the FTs in NLOS environments. The TOA measurements are multiplied by the speed of light to obtain range estimates which can be used together with trilateration techniques to determine the position of the UE. Applications arise in locating emergency responders or avalanche/earthquake victims as well as locating valuable items in manufacturing plants [37].
Recall from Section 1.1 that the NLOS error statistics for TOA range measurements are positively biased. This makes conventional least-squares (LS) techniques lose in positioning accuracy. Thus, there is a demand for robust alternatives.
In this chapter, the nonlinear relationship between the range estimates and the position of the UE is linearized and robust and semi-parametric techniques are used to estimate the position of the UE. A semi-parametric estimator for asymmetric error statistics together with results from [43, 45] are presented.

3.1 Problem Statement

Consider a UE surrounded by M FTs. The UE does not move within a certain time frame $\Delta t K$ seconds where Δt is the sampling period and K is the number of time samples, i.e., the number of TOA estimates. After multiplying the TOA estimates by the speed of light the general signal model (2.1) from Section 2.1 can be written for a discrete time step k as

$$y_m(k) = \underbrace{\sqrt{(x - x_{\text{FT},m})^2 + (y - y_{\text{FT},m})^2}}_{=h_m(\boldsymbol{\theta})} + v_m(k) \quad m = 1, 2, \ldots, M, \quad (3.1)$$

where $h_m(\cdot)$ describes the Euclidean distance between the m-th FT and the UE. For simplicity we assume that $h_m(\boldsymbol{\theta})$ does not vary much for different FTs, meaning the distances from the UE to the different FTs are similar. Sensor noise and perturbations due to NLOS effects are modeled as iid random quantities $v_m(k)$ according to noise model (2.2) ($f_V(v) = (1-\varepsilon)\mathcal{N}(v; 0, \sigma_G^2) + \varepsilon \mathcal{H}(v)$, with $\mathcal{H}(v) = \mathcal{N}(v; 0, \sigma_G^2) * f_\eta$). LOS propagation, resulting in zero-mean Gaussian sensor noise, occurs with probability $1-\varepsilon$. The sensor noise variance σ_G^2 is assumed to be known. In contrast NLOS propagation occurs with probability ε and the unknown pdf f_η describes the positive errors due to NLOS [92]. Different models for f_η such as an exponential, a shifted Gaussian and Rayleigh pdf are available in the literature [41, 53, 73, 111]. Thus, the density $f_V(v)$ has a positive mean unless $\varepsilon = 0$.

3.1.1 State of the Art

Assume that k is fixed and no prior knowledge about the position of the UE is available. Then, in LOS, minimizing the sum of the squared residuals (2.4) yields the optimal solution. However, since $h_m(\cdot)$ is nonlinear, even a quadratic cost function (as in least-squares estimation) can result in several local minima and local search algorithms [31], e.g. the Gauss-Newton algorithm [11] can lead to biased estimates. In other words, Equation (2.5) from Section 2.1.2 does not have a unique solution which requires good initialization for the local search algorithms. These algorithms iteratively determine the location of the UE based on a first-order approximation of the measurement model (3.1). However, outliers due to NLOS propagation or a starting point too far away from the true position can result in convergence problems and position accuracy can substantially be reduced.

One way to cope with the NLOS problem is to identify the NLOS FTs and use the remaining ones for positioning based on the above mentioned methods. NLOS detection algorithms are proposed in [10, 36, 83, 101] and [91] provides an overview of NLOS identification techniques. The approaches developed in [10, 83] are based on hypothesis testing where a Gaussian pdf for the NLOS error is assumed. The authors in [101] use more general statistical LOS/NLOS identification schemes where outlier detection is performed based on higher order moments or tests for Gaussianity. A statistical non-parametric NLOS detection approach is suggested in [36] where the error pdf of the observations from one FT is estimated using KDE and statistical distance measures are used to compare the pdf estimate with the Gaussian pdf. This approach requires several observations per FT for performing KDE and a FT is accepted to be LOS if the distance of its empirical error pdf to the Gaussian pdf is smaller than a certain threshold. However, the concept of detecting the NLOS FTs and discarding them for positioning has two major limitations. If a mis-detection occurs, higher positioning errors are expected using standard LS techniques because outliers have a deleterious effect on these estimates. In contrast, a false alarm reduces the set of LOS FT which increases the positioning error as well since less observations are left for estimation. In general, problems with this approach occur when a small amount of FTs remain (e.g. a number of FTs smaller than three, leading to ambiguities) or when the remaining FTs are spaced in a disadvantageous geometric constellation, e.g., when all remaining FTs lay approximately on a straight line. Therefore, we concentrate on approaches that use

3.1 Problem Statement

the observations from all FTs for positioning.

In [18, 22, 40] NLOS mitigation algorithms based on TOA measurements are proposed. The three approaches consider grouping of range measurements to obtain an LS position estimate from each subgroup. The position estimates are then combined in different manners so that deleterious estimates are underweighed with respect to accurate position estimates. All references mentioned above avoid imposing a specific NLOS error distribution, which is convenient since it is unknown in practice. Chen [22] suggests an NLOS mitigation algorithm based on residual weighting. The observations are formed into different subgroups and LS estimation is used for each group to determine the position of the UE and its residual error (the sum of the squared residuals from the obtained position estimate). The overall position estimate is a weighted combination of the different estimates where the weights are determined as the inverse of the residual error, i.e., high residual errors result in small weights and low residual errors result in larger weights. Since the final position estimate is the weighted sum of the different position estimates, an error in the residual weights can have a large impact on the final position estimate.

Apart from the grouping, which is done to obtain $\binom{M}{3}$ subgroups, the authors in [18] suggest a different approach to overcome this problem. For each position estimate from a subgroup the *median* of the squared residuals is calculated. Then the final solution is the position estimate corresponding to the minimum of the *medians* from each subgroup. Taking the *median* (instead of the *mean* which corresponds to the sum in [22]) ensures that the position estimates corresponding to very large or very small squared residuals are not taken into account. This is because the *median* is the number that separates the higher half of the sample from the lower half which is achieved by ordering the data.

However, the performance of both algorithms [18, 22] strongly depends on the optimization algorithms used for each subgroup to obtain the LS position estimate. While local search algorithms can fail when outliers occur or initialization is erroneous, more sophisticated and thus more complex optimization tools [11] rather find the global minimum instead of falsely selecting a local one. Given that the latter techniques are available, one could also use a different penalty function to decrease the impact of NLOS outliers and search for the global minimum to obtain robust position estimates. That means one can use a penalty function that is increasing less severely than a quadratic one, e.g., $\rho_{c_1}(v)$ in Equation (2.16) is increasing linearly beyond a certain threshold, such an approach is applied to geolocation in [72].

An alternative is to estimate the penalty function non-parametrically and determine the position of the UE by searching for the global minimum or minimizing a non-parametric estimate of the entropy of the residuals [98], as in [110] (semi-parametric approaches). Another alternative is to use constraint optimization techniques [106] that take into account that the NLOS error is always positive and thus search for the minimum in a restricted area. However, in many applications computational power is limited and therefore numerically simpler solutions are preferred and considered in this thesis.

To prevent convergence problems, when using local search algorithms, the signal model is linearized to obtain a closed-form solution and robust and semi-parametric regression techniques for linear models are applied.

3.1.2 Linearization

For any estimator, it is desirable that the number of observations is large to exploit as much information as possible. This is in particular true for semi-parametric estimators because non-parametric KDE is used to estimate the noise pdf. Thus, a linearization that maintains the number of observations in the linearized model is preferred.

In [89] several LS closed-form solutions based on linearization of the signal parameters TOA, AOA, TDOA and RSS are derived. The explicit solution for TOA measurements for a fixed time step k combines M range measurements such that $M-1$ observations are left for an LS estimator in the linearized model. Considering K measurements per FT as in (3.1) gives us $K(M-1)$ samples for determining the position of the UE. In contrast, for a fixed k, M samples after linearization are used in [23, 106] to determine the UE position. This approach is preferable because of the larger amount of observations. It is followed here and extended to K range measurements per FT resulting in a total of KM observations.

Let the parameter of interest be $\boldsymbol{\theta} = [x\ y\ R]^\mathsf{T}$ where $R = x^2 + y^2$. For the m-th FT we can write the k-th range measurement in the noise-free case as

$$\begin{aligned} h_m^2(\boldsymbol{\theta}) &= (x_{\mathsf{FT},m} - x)^2 + (y_{\mathsf{FT},m} - y)^2 \quad m = 1, 2, \ldots, M \quad (3.2)\\ &= x_{\mathsf{FT},m}^2 + y_{\mathsf{FT},m}^2 - 2x_{\mathsf{FT},m}x - 2y_{\mathsf{FT},m}y + R. \quad (3.3) \end{aligned}$$

In practice, the measurements are contaminated by Gaussian sensor noise and NLOS perturbations as in (3.1). Squaring (3.1) yields

$$\begin{aligned} y_m^2(k) &= h_m^2(\boldsymbol{\theta}) + \tilde{v}_m(k) \quad m = 1, 2, \ldots, M,\ k = 1, 2, \ldots, K, \quad (3.4)\\ \tilde{v}_m(k) &= 2h_m(\boldsymbol{\theta})(g_m(k) + \eta_m(k)) + (g_m(k) + \eta_m(k))^2, \end{aligned}$$

where $g_m(k)$ are realizations of a Gaussian random variable modeling sensor noise and $\eta_m(k)$ are realizations of a positive random variable modeling perturbations due to the NLOS effect. To model the pdf $f_{\tilde{V}}(\tilde{v})$ of the random nonlinearities $\tilde{\mathbf{v}}_m(k)$ requires knowledge of the true position $\boldsymbol{\theta}$ and the NLOS error density f_η. Both quantities are unknown in reality and we therefore leave $f_{\tilde{V}}(\tilde{v})$ unspecified in the construction of the estimator (thus taking a semi-parametric approach). Note that even in the nominal case, where no NLOS components are present, $f_{\tilde{V}}(\tilde{v})$ can have an asymmetric shape if $h_m(\boldsymbol{\theta})$ is sufficiently small because of the linear combination $2h_m(\boldsymbol{\theta})g_m(k) + g_m^2(k)$ of a Gaussian random variable and a squared Gaussian random variable. The more NLOS components are included in (3.4), i.e, the higher the probability of NLOS occurrences ε, the more right skewed $f_{\tilde{V}}(\tilde{v})$ becomes.

For M FTs and K measurements, Equation (3.4) can be rewritten into the linear regression model

$$\tilde{\mathbf{y}} = \mathbf{D}\boldsymbol{\theta} + \tilde{\mathbf{v}}, \tag{3.5}$$

where

$$\tilde{\mathbf{y}} = \begin{bmatrix} y_1^2(1) - (x_{\mathsf{FT},1}^2 + y_{\mathsf{FT},1}^2) \\ y_1^2(2) - (x_{\mathsf{FT},1}^2 + y_{\mathsf{FT},1}^2) \\ \vdots \\ y_1^2(K) - (x_{\mathsf{FT},1}^2 + y_{\mathsf{FT},1}^2) \\ \vdots \\ y_M^2(1) - (x_{\mathsf{FT},M}^2 + y_{\mathsf{FT},M}^2) \\ \vdots \\ y_M^2(K) - (x_{\mathsf{FT},M}^2 + y_{\mathsf{FT},M}^2) \end{bmatrix}, \quad \mathbf{D} = \begin{bmatrix} -2x_{\mathsf{FT},1} & -2y_{\mathsf{FT},1} & 1 \\ -2x_{\mathsf{FT},1} & -2y_{\mathsf{FT},1} & 1 \\ \vdots & \vdots & \vdots \\ -2x_{\mathsf{FT},1} & -2y_{\mathsf{FT},1} & 1 \\ \vdots & \vdots & \vdots \\ -2x_{\mathsf{FT},M} & -2y_{\mathsf{FT},M} & 1 \\ \vdots & \vdots & \vdots \\ -2x_{\mathsf{FT},M} & -2y_{\mathsf{FT},M} & 1 \end{bmatrix},$$

and $\mathbf{D}$ is termed regressor matrix. The vector $\tilde{\mathbf{v}}$ contains the elements $\tilde{v}_m(k)$ from (3.4) with pdf $f_{\tilde{V}}(\tilde{v})$. Note that $\dim(\tilde{\mathbf{y}}) = M \cdot K \times 1$ and $\dim(\mathbf{D}) = M \cdot K \times 3$. The linear model in (3.5) allows us to use an explicit solution, which is shown in Section 3.2.

3.2 Approaches for Position Estimation

In this section, different statistical approaches to estimate the position of the UE are explained and the corresponding algorithms are presented. We start with parametric ML, least-squares and robust methods and finally suggest a novel semi-parametric estimator for asymmetric noise environments [43, 45].

3.2.1 Maximum Likelihood and Least-Squares Estimation

Assume that $\tilde{\mathbf{v}}$ is iid (which can only hold if $h_1(\boldsymbol{\theta}) = h_2(\boldsymbol{\theta}) = \ldots = h_M(\boldsymbol{\theta})$) with known pdf $f_{\tilde{V}}(\tilde{v})$, then the MLE is

$$\hat{\boldsymbol{\theta}}_{\mathsf{ML}} = \arg\min_{\boldsymbol{\theta}} \sum_{i=1}^{MK} -\log f_{\tilde{V}}\left(\tilde{y}_i - \sum_{j=1}^{\dim(\boldsymbol{\theta})} [\mathbf{D}]_{ij}\theta_j\right), \tag{3.6}$$

where $[\mathbf{D}]_{ij}$ denotes the j-th element of the i-th row of matrix $\mathbf{D}$. Solving the derivative of (3.6) for zero yields

$$\sum_{i=1}^{MK} [\mathbf{D}]_{ij} \varphi \left(y_i - \sum_{j'=1}^{\dim(\boldsymbol{\theta})} [\mathbf{D}]_{ij'} \theta_{j'} \right) = 0, \quad j = 1, \ldots, \dim(\boldsymbol{\theta}). \tag{3.7}$$

If $f_{\tilde{V}}(\tilde{v})$ is Gaussian (3.7) comes down to the LS solution, i.e.,

$$\hat{\boldsymbol{\theta}}_{\mathsf{LS}} = (\mathbf{D}^{\mathsf{T}}\mathbf{D})^{-1}\mathbf{D}^{\mathsf{T}}\tilde{\mathbf{y}}. \tag{3.8}$$

However, even for $\varepsilon = 0$, Equation (3.8) does not produce the MLE because the Gaussian assumption of $f_{\tilde{V}}(\tilde{v})$ is not fulfilled. This is due to the squared Gaussian random variable in (3.4) and the noise sample $\tilde{v}_m(k)$ depending on the true Euclidean distances $h_m(\boldsymbol{\theta})$ between the UE and the FTs. Even though we assume that $h_m(\boldsymbol{\theta})$ does not change much between different FTs, small variations in $h_m(\boldsymbol{\theta})$ can emphasize or reduce the impact of the noise sample $\tilde{v}_m(k)$. Thus, it becomes desirable to downweigh the observations with larger noise scales and emphasize the contribution of observations with smaller noise scales. This can be achieved by taking the weighted least-squares (WLS) estimator

$$\hat{\boldsymbol{\theta}}_{\mathsf{WLS}} = \arg\min_{\boldsymbol{\theta}} (\tilde{\mathbf{y}} - \mathbf{D}\boldsymbol{\theta})^{\mathsf{T}} \mathbf{W}^{-1} (\tilde{\mathbf{y}} - \mathbf{D}\boldsymbol{\theta}), \tag{3.9}$$

yielding

$$\hat{\boldsymbol{\theta}}_{\mathsf{WLS}} = (\mathbf{D}^{\mathsf{T}} \mathbf{W}^{-1} \mathbf{D})^{-1} \mathbf{D}^{\mathsf{T}} \mathbf{W}^{-1} \tilde{\mathbf{y}}, \tag{3.10}$$

where $\mathbf{W} \in \mathbb{R}^{KM \times KM}$ is a weighting matrix that copes with the different noise scales mentioned above. If the noise is uncorrelated among FTs and time, which is assumed here, $\mathbf{W}$ is diagonal. For LOS, in [23] the covariance matrix $\mathbf{W}$ for high SNRs conditions is determined based on approximating the noise in (3.4) as

$$\tilde{v}_m(k) \approx 2h_m(\boldsymbol{\theta}) g_m(k). \tag{3.11}$$

Then, the variance of each observation noise sample is

$$\mathsf{E}\{\tilde{v}_m^2(k)\} = 4h_m^2(\boldsymbol{\theta})\sigma_G^2 \quad \forall \; k = 1, 2, \ldots, K, \; m = 1, 2, \ldots, M, \tag{3.12}$$

yielding the following covariance matrix

$$\begin{aligned}\mathbf{W}^* &= \mathsf{diag}[\mathsf{E}\{\tilde{v}_1^2(1)\}, \mathsf{E}\{\tilde{v}_1^2(2)\}, \ldots, \mathsf{E}\{\tilde{v}_1^2(K)\}, \ldots, \\ &\quad \mathsf{E}\{\tilde{v}_M^2(1)\}, \ldots, \mathsf{E}\{\tilde{v}_M^2(K)\}].\end{aligned} \tag{3.13}$$

In practice the true distances $h_m(\boldsymbol{\theta})$ are not known which makes computation of $\mathbf{W}^*$ infeasible. The authors in [23, 106] suggest to take the observations $y_m(k)$ instead of $h_m(\boldsymbol{\theta})$ to approximate the true covariance matrix yielding $\mathbf{W}$. Then, (3.10) can be used to obtain the WLS estimates which serves as a reference for comparison purposes.

3.2 Approaches for Position Estimation

However, in NLOS environments, $f_{\tilde{V}}(\tilde{v})$ strongly deviates from Gaussianity and approximation (3.11) does not hold anymore leading to a loss of positioning accuracy when using LS (3.8) or WLS (3.10) estimators. Thus, robust estimation techniques are required to limit the impact of NLOS outliers. Note that performance can be improved when the relationship $R = x^2 + y^2$ is incorporated into the estimators [23]. This step is left out here for all methods since our aim is to investigate the sensitivity towards NLOS outliers.

3.2.2 Robust M-estimation

Since NLOS outliers significantly decrease performance of conventional ML or LS estimators, the score function $\varphi(\cdot)$ in (3.7) is replaced by a bounded antisymmetric score function $\psi(\cdot)$ that limits the effect of NLOS outliers to obtain robust position estimates. Recall that the exact minimax solution, given in Section 2.1.3.1, minimizes the maximum asymptotic variance of the position estimates for the least favorable distribution and requires knowledge of ε and the standard deviation of the nominal noise model $\sigma_{\tilde{G}}$. Even though $\sigma_{\tilde{G}}$ can be approximated by using the observations $\tilde{\mathbf{y}}$ in (3.12), ε is always unknown which makes the exact minimax estimator inapplicable. Instead, an approximate solution, introduced in [105], can be used to determine the position of the UE. The normalized score function of this estimator is

$$\psi_{c_1}(v) = \begin{cases} v, & |v| \leq c_1 \\ c_1 \operatorname{sign}(v), & |v| > c_1. \end{cases} \qquad (3.14)$$

where c_1 is a clipping point that can be chosen in terms of the efficiency loss we are willing to pay under the nominal assumption. For this purpose, the standard deviation of the noise $\sigma_{\tilde{V}}$ is estimated. Then, the residuals, obtained from an LS estimate, are divided by $\sigma_{\tilde{V}}$ to fit the score function (3.14) which replaces $\varphi(\cdot)$ in (3.7) to estimate $\boldsymbol{\theta}$. This ensures that large outliers are clipped whereas the "good" measurements are exploited for estimation.

However, since $f_{\tilde{V}}(\tilde{v})$ is asymmetric with a positive mean the symmetric assumption for the soft-limiter is not fulfilled anymore resulting in biased position estimates. To cope with deviations from symmetry, in [24, 49], a redescending score function $\psi_{c_1,c_2}(v)$ which goes down to zero beyond a second threshold c_2 is suggested. It is depicted in Figure 2.1(d) in Section 2.1.3.1 and its mathematical expression is

$$\psi_{c_1,c_2}(v) = \begin{cases} v, & |v| \leq c_1 \\ b \tanh(0.5b(c_2 - |v|)) \operatorname{sign}(v), & c_1 < |v| \leq c_2, \\ 0 & |v| > c_2 \end{cases} \qquad (3.15)$$

where b is chosen such that $\psi_{c_1,c_2}(v)$ is continuous at c_1. This function discards large outliers completely and can therefore obtain unbiased estimates when $f_{\tilde{V}}(\tilde{v})$ is asymmetric. Note that the estimator based on $\psi_{c_1,c_2}(v)$ requires that $f_{\tilde{V}}(\tilde{v})$ is symmetric within the central region $[-c_1, c_1]$ which is only fulfilled here if (3.11) holds. Otherwise a model mismatch occurs which decreases performance of the robust estimator. The score function (3.15) is used throughout this work since it leads to unbiased estimates in asymmetric noise environments [24, 49].

Note that other bounded score functions with one or various tuning parameters exist in the literature [49, 51]. However, one problem with redescending score functions in general is the choice of their clipping parameters. Several values of the clipping parameters can lead to the same efficiency loss in the Gaussian case. Thus the question arises which set of values are best for a specific situation. Furthermore, if the clipping parameters are too small a high efficiency loss is expected in the Gaussian case which can even result in convergence problems when ε increases. In contrast, if the clipping parameters are chosen too large, the estimator loses robustness in NLOS environments. In addition to that all parametric robust estimators share the same shortcoming that they can only slightly adapt to the noise scenario by tuning their clipping parameters. Hence, they are suboptimal for any other noise distribution than the specified one.

These limitations can be overcome by using semi-parametric estimators which are less restrictive than their parametric counterparts because less assumptions are made on the noise environment. The noise pdf $f_{\tilde{V}}(\tilde{v})$ is estimated non-parametrically and is used to obtain the position estimate based on the ML principle. Since parametric M-estimators are a classical tool to cope with deviation from Gaussianity they are used for comparison purposes later on.

The algorithms to compute the position estimate for a particular parametric score function $\psi(\cdot)$ (either $\psi_{c_1}(v)$ or $\psi_{c_1,c_2}(v)$) are depicted in Figure 3.1 [51]. First, at Step 1), an initial position estimate $\hat{\boldsymbol{\theta}}_0$ is obtained, e.g. via LS. This estimate is taken to determine the noise residuals $\hat{\mathbf{v}}$ at Step 2). Then, at Step 3) the standard deviation of the noise is estimated using the mad explained in Section 2.1.3.2. At Step 4), there are two possibilities to update the parameter estimate: One can either compute a Newton-Raphson step which is termed "*modified residuals*" where the new parameter estimate is the previous one plus the pseudoinverse multiplied by the score function evaluated at the normalized residuals. The alternative is to do the update by a WLS estimator termed "*modified weights*" where the weights are computed as the ratio of the score function evaluated at the normalized residuals and the normalized residuals. In Step 5) the algorithm checks for convergence, meaning if the norm of the previous and actual position estimate is smaller than the required precision the algorithms stops. Otherwise, we go to Step 2) to repeat the different steps until convergence is achieved.

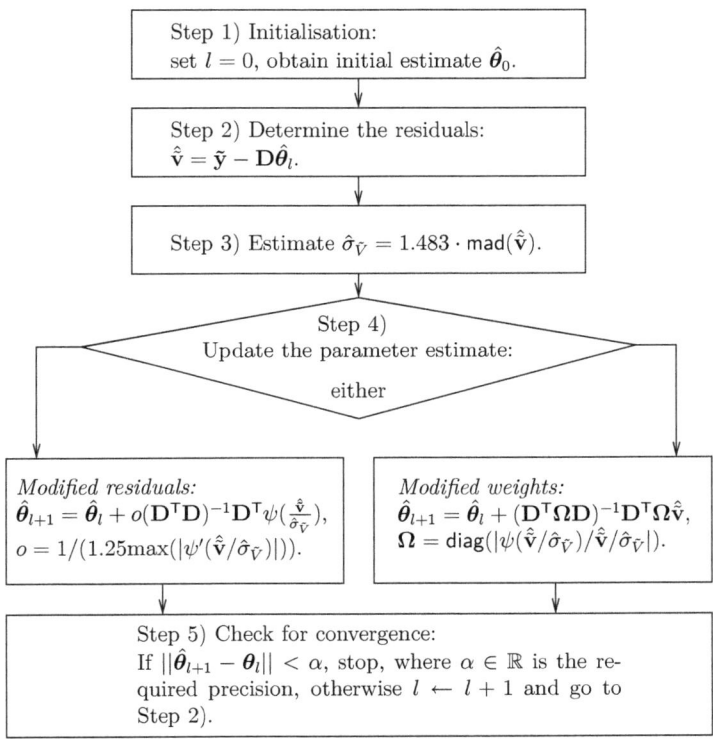

Figure 3.1: Robust M-estimation algorithm [51].

3.2.3 Positioning Based on Semi-Parametric Estimation

3.2.3.1 General Concept

Instead of using a parametric score function for estimating $\boldsymbol{\theta}$, the score function is calculated from non-parametric estimates of the noise pdf to estimate the position of the UE based on the idea of semi-parametric statistics explained in Section 2.1.4.2. To do so we use the residuals

$$\hat{\tilde{\mathbf{v}}} = \tilde{\mathbf{y}} - \mathbf{D}\hat{\boldsymbol{\theta}}_0, \qquad (3.16)$$

where $\hat{\boldsymbol{\theta}}_0$ is an initial parameter estimate (obtained, e.g., via LS), to determine $f_{\tilde{V}}(\tilde{v})$ and its derivative $f'_{\tilde{V}}(\tilde{v})$ non-parametrically. This allows us to incorporate information about the noise pdf into the estimator. The score function estimate is then

$$\hat{\varphi}(\tilde{v}) = -\hat{f}'_{\tilde{V}}(\tilde{v})/\hat{f}_{\tilde{V}}(\tilde{v}) \qquad (3.17)$$

and is used in (3.7) to replace $\varphi(\cdot)$ for determining $\hat{\boldsymbol{\theta}}$. Details on the implementation are given in Section 3.2.3.4. We assume that $\tilde{\mathbf{v}}$ are independent but we do not restrict ourselves to identically distributed data. Furthermore, it is assumed that the variance among $\tilde{v}_m \ \forall m = 1, 2, \ldots, M$ does not vary much which holds if the variability of $h_m(\boldsymbol{\theta})$ is small.

Theoretical convergence and optimality in the Fisher sense of the semi-parametric estimator from (3.17) and (3.7) (without transformation density) is established, e.g., in [8] for linear regression models, under different sets of assumptions. From a statistical viewpoint, one may argue that the KDE step requires larger sample sizes (than those considered here) in order to be accurate. In practice, however, it turns out that high accuracy of the non-parametric density or score function estimate (3.17) is not needed for the location estimator to perform well. One should recall that the aim is to estimate the global optimum, i.e., the zero of the score function in the estimation equation (3.7), and not the actual pdf of the residuals.

3.2.3.2 Transformation KDE for Asymmetric Noise Densities

Since outliers occur in the residuals due to NLOS effects, using conventional KDE as in (2.26) fails because it leads to multiple maxima at the tails of the distribution, which produce ambiguities in (3.7). This effect is described in more detail in Section 2.1.4.2. However, unlike in Section 2.1.4.2 where a symmetric density $f_V(v)$ is considered, here $f_V(v)$ becomes asymmetric due to the positive bias introduced by the NLOS effects for TOA geolocation. The asymmetric property of the density $f_V(v)$ is inherited by the density of the linearized model $f_{\tilde{V}}(\tilde{v})$ which is light-tailed on the left side and heavy-tailed on the right side. Thus, positive outliers occur in the residuals $\hat{\tilde{\mathbf{v}}}$ which even complicate the estimation of $\hat{f}_{\tilde{V}}(\tilde{v})$. This is because symmetrizing the residuals, which is done for symmetric samples in Section 2.1.4.2 to smooth the pdf estimate and to improve small sample performance, yields to a model mismatch of $f_{\tilde{V}}(\tilde{v})$ and consequently to erroneous position estimates $\hat{\boldsymbol{\theta}}$. Hence, transforming the residuals with a antisymmetric function around the origin, e.g., see the modulus transformation in Figure 2.4(b), is not appropriate since it only yields to a rescaled but asymmetric sample in the transformed domain.

3.2 Approaches for Position Estimation

If the prior knowledge of asymmetry is incorporated into the estimator, better performance can be expected. Note that in some cases, e.g., when no NLOS components are contained in the observations, $f_{\tilde{V}}(\tilde{v})$ can remain symmetric when approximation (3.11) holds. Thus, an estimator is required that takes into account the degree of asymmetry in the underlying data automatically.

For this purpose, we transform the residuals using a transformation $t(\hat{\mathbf{v}}, \lambda)$ given in [112], where λ controls the shape of this function. If this parameter is chosen appropriately the function $t(\hat{\mathbf{v}}, \lambda)$ suits asymmetric data such that the data after transformation is *almost* Gaussian [112]. Hence we assume a symmetric, but non-Gaussian pdf in the transformed domain in order to remain adaptive. The expression for $t(\hat{\mathbf{v}}, \lambda)$ and a selection scheme for λ are given later. We symmetrize the transformed residuals into

$$\mathbf{w}_s = [-t(\hat{\mathbf{v}}, \lambda), +t(\hat{\mathbf{v}}, \lambda)], \qquad (3.18)$$

in order to implement the constraint that the pdf is symmetric which doubles the sample size and consequently improves small sample performance. The density estimate and its derivative in the original domain is

$$\hat{f}_{\tilde{V}}(\tilde{v}) = \frac{1}{\dim(\mathbf{w}_s)\delta} \sum_{i=1}^{\dim(\mathbf{w}_s)} \mathcal{K}\left(\frac{t(\tilde{v}, \lambda) - w_i}{\delta}\right) \left(\left|\frac{dt^{-1}(w, \lambda)}{dw}\right|\right)^{-1}, \qquad (3.19)$$

$$\hat{f}'_{\tilde{V}}(\tilde{v}) = -\frac{1}{\dim(\mathbf{w}_s)\delta} \sum_{i=1}^{\dim(\mathbf{w}_s)} \mathcal{K}\left(\frac{t(\tilde{v}, \lambda) - w_i}{\delta}\right) \left(\frac{t(\tilde{v}, \lambda) - w_i}{\delta^2}\right)$$
$$\cdot \frac{dt(\tilde{v}, \lambda)}{d\tilde{v}} \left(\left|\frac{dt^{-1}(w, \lambda)}{dw}\right|\right)^{-1}, \qquad (3.20)$$

where $w_i = t(\tilde{v}_i, \lambda)$ and $\dim(\mathbf{w}_s) = 2MK$ for the geolocation problem. These formulas are plugged into (3.17) and (3.7) to estimate $\boldsymbol{\theta}$. A function that transforms the residuals in a symmetric, *approximately* Gaussian sample is [112]

$$t(v, \lambda) = \begin{cases} [(|v|+1)^{\lambda} - 1]/\lambda, & \lambda \neq 0, \ v \geq 0 \\ \log(v+1), & \lambda = 0, \ v \geq 0 \\ -[(-v+1)^{2-\lambda} - 1]/(2-\lambda), & \lambda \neq 2, \ v < 0 \\ -\log(-v+1), & \lambda = 2, \ v < 0. \end{cases} \qquad (3.21)$$

Unlike the modulus transformation [57], used in [44] to reduce kurtosis of the residuals, transformation (3.21) is appropriate for reducing kurtosis and skewness of a given data set [112]. This function is depicted in Figure 3.2. Note that $t(v, \lambda)$ is concave for $\lambda < 1$, convex for $\lambda > 1$ and linear for $\lambda = 1$. In particular positive outliers due to NLOS propagation are brought closer together while negative ones are projected further apart when $\lambda < 1$ which is desirable for the TOA geolocation problem. A discussion on how to select the shape parameter λ and the global bandwidth δ is given in Section 3.2.3.3.

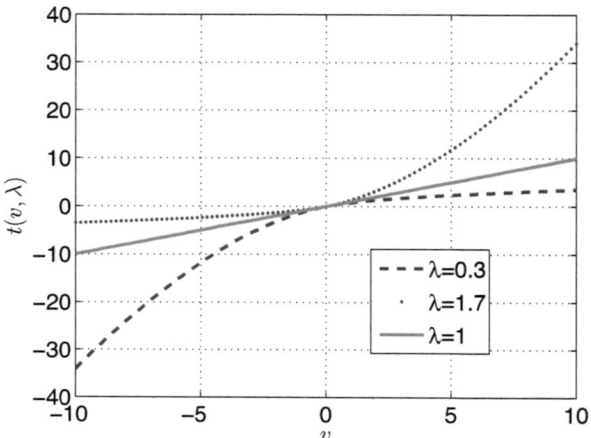

Figure 3.2: Transformation function $t(v, \lambda)$ for different values of λ.

3.2.3.3 Selection of the Tuning Parameters δ and λ

For the selection of the global bandwidth δ in (3.19) or (3.20), many automatic selection rules can be found in the literature, see, e.g., [7, 27, 93]. Some procedures optimize a certain criterion and are computationally demanding. Other alternatives provide a bandwidth instantaneously and prove to be sufficient in many problems [7, 93]. An example of such a selection scheme is to estimate δ by $\hat{\delta} = 1.06\hat{\sigma}_W(\mathsf{dim}(\mathbf{w}))^{-1/5}$ [93], where $\hat{\sigma}_W = 1.483\mathsf{mad}[t(\hat{\mathbf{v}}, \lambda)]$. This bandwidth selection rule is used in the numerical studies presented here, where mad is the mean absolute deviation. The next question that arises is how to choose λ in order to achieve a pdf in the transformed domain that is as close as possible to Gaussianity. Following [57, 112], we assume (only for selection of λ) that $f_W(w)$ is a Gaussian pdf with mean μ_W and variance σ_W^2 that can be estimated using the sample mean and variance of the transformed data $\mathbf{w}$. When we assume iid residuals, the likelihood function in the original domain is constructed by taking the product of the pdfs $f_{\tilde{V}}(\hat{\tilde{v}}_i)$. The MLE of λ is then obtained by maximizing the log-likelihood function, i.e.,

$$\hat{\lambda} = \arg\max_{\lambda} \left\{ -\frac{\mathsf{dim}(\tilde{\mathbf{v}})}{2} \log \hat{\sigma}_W^2(\lambda) + (\lambda - 1) \sum_{i=1}^{\mathsf{dim}(\tilde{\mathbf{v}})} \mathsf{sign}(\hat{\tilde{v}}_i) \log(|\hat{\tilde{v}}_i| + 1) \right\}, \quad (3.22)$$

3.2 Approaches for Position Estimation

Unlike in [44], it is preferable here to restrict λ to the (left) neighborhood of one. This is because the sample of residuals $\hat{\mathbf{v}}$ has to be transformed such that its right skewness is corrected for and such that the sample after transformation is symmetric. This can only be fulfilled when $\lambda < 1$, for which the (heavier) right tail of $\hat{f}_{\tilde{V}}(\tilde{v})$ is transformed closer to the *median* of the sample, and the (lighter) left tail becomes heavier. Since the transformation function $t(v, \lambda)$ (3.21) is not antisymmetric, choosing λ too far below one can scatter the residuals of the left tail too far apart. Therefore λ should not be chosen too small compared to one. An example of the impact of λ on geolocation is given in Section 3.3, Figure 3.5.

3.2.3.4 Algorithm

We present two algorithms, illustrated in Figure 3.3, to compute the position estimates $\hat{\boldsymbol{\theta}}$ based on the principle of semi-parametric statistics. The algorithms are similar to the ones for parametric M-estimation except that the score function is estimated non-parametrically. After Step 1), where an initial position estimates is obtained via LS, the residuals are determined in Step 2). Then, at Step 3), the score function is estimated non-parametrically. Unlike in the algorithm for parametric M-estimation, where the standard deviation is estimated as a nuisance parameter for normalizing the residuals, here, the whole pdf is the nuisance parameter which leads to the score function estimate $\hat{\varphi}(\cdot)$. Estimation of the score function first involves estimation of λ using (3.22) and transformation of the residuals with (3.21) into $\mathbf{w} = t(\hat{\mathbf{v}}, \hat{\lambda})$. Next, estimates of $f_{\tilde{V}}(\tilde{v})$ and its derivative are obtained using (3.19) and (3.20), respectively. The update, in Step 4), can either be performed using a Newton-Raphson or a WLS approach. Convergence is checked in Step 5) and the algorithm stops, as soon as the required precision α is achieved.

To save computational power, unlike in Figure 2.3(a), the pdf estimate and consequently the estimated score function are only evaluated at the residuals $\hat{\mathbf{v}}$. Note that the pseudoinverse $(\mathbf{D}^\mathsf{T}\mathbf{D})^{-1}\mathbf{D}^\mathsf{T}$ can be computed offline which seem to make *modified residuals* more favorable in terms of computation complexity with respect to *modified weights*. However, in contrast the algorithms based on *modified weights* requires less iterations to achieve convergence. Since their performance is similar [45] only the algorithm based on *modified residuals* is used in the sequel.

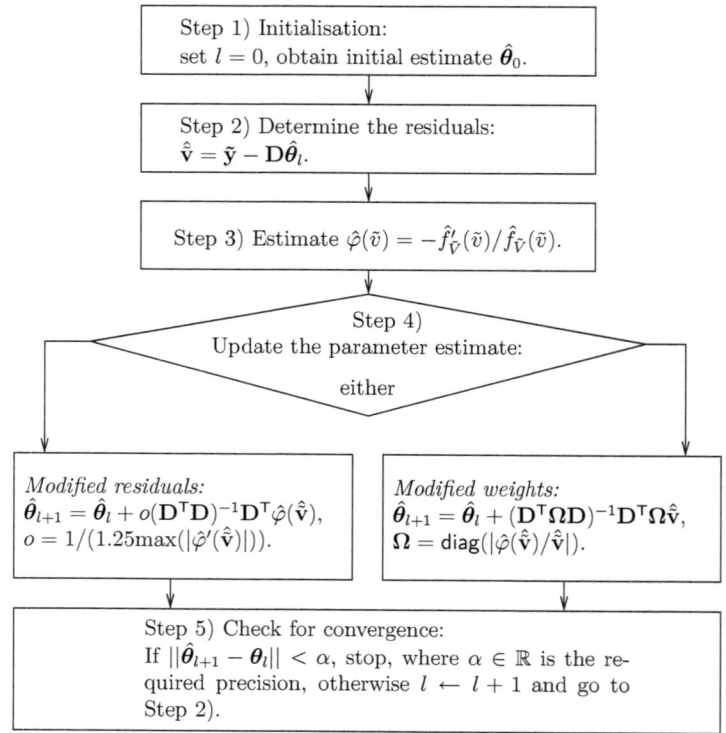

Figure 3.3: Semi-parametric algorithms.

3.3 Numerical Study

3.3.1 Simulation Environments and Settings

Consider a UE surrounded by $M = 10$ FTs in a 2D-plane depicted in Figure 3.4 where the x- and y-position of the UE for each Monte-Carlo run is uniformly distributed between 2 and 3km. It is estimated using the standard LS estimator derived in Section 3.2.1 in (3.8) and the WLS estimator [23] given in the same section labeled as LS and WLS, respectively. Furthermore we use the M-estimator from Section 3.2.2 based on the redescending score function (3.15) using the algorithm in Figure 3.1. Since the performance of the *modified weights* and *modified residuals* algorithms is similar [51]

3.3 Numerical Study

we arbitrarily choose the one based on *modified residuals* labeled as as R-MR (*robust modified residuals*). The clipping parameters of the redescending score function are chosen as $c_1 = 1.5$ and $c_2 = 2.5$ to achieve a trade-off between efficiency loss in the Gaussian case and robustness for the non-Gaussian case. A numerical study on the choice of the two clipping parameters for LOS and NLOS scenarios can be found in Appendix A.1.

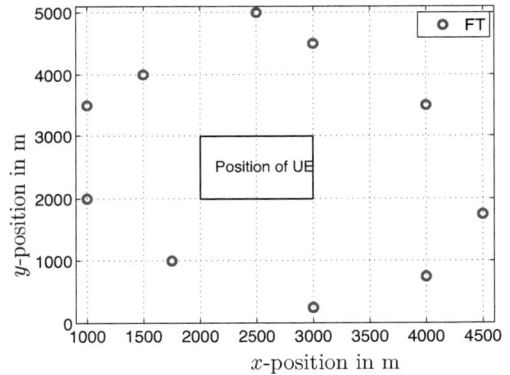

Figure 3.4: Network with $M = 10$ FTs. The position of the UE is uniformly distributed in the rectangle. The positions of the FTs are $(x_1 = 2.5\text{km}, y_1 = 5\text{km})$, $(x_2 = 1\text{km}, y_2 = 3.5\text{km})$, $(x_3 = 4.5\text{km}, y_3 = 1.75\text{km})$, $(x_4 = 1.5\text{km}, y_4 = 4\text{km})$, $(x_5 = 3\text{km}, y_5 = 4.5\text{km})$, $(x_6 = 1.75\text{km}, y_6 = 1\text{km})$, $(x_7 = 4\text{km}, y_7 = .75\text{km})$, $(x_8 = 4\text{km}, y_8 = 3.5\text{km})$, $(x_9 = 1\text{km}, y_9 = 2\text{km})$, $(x_{10} = 3\text{km}, y_{10} = .25\text{km})$.

The M-estimator using the soft-limiter in (3.14) achieves similar performance as the estimator based on the redescending score function (3.15) used throughout the simulation studies. It is numerically more stable but becomes biased as ε increases. For illustration purposes the mean absolute deviation is used for scale estimation.

For the semi-parametric estimator we choose the *modified residuals* algorithm depicted in Figure 3.3, labeled as SP-MR. The break condition for R-MR and SP-MR for convergence is arbitrarily set to $\alpha = 10^{-3}$m and a maximum number of 20 iterations are used. The algorithms [18, 22] that do not impose any NLOS error pdf are appealing for comparison purposes. We propose to use the method in [18] which is based on the *least median of squares* [86] from *robust statistics*. It has a breakdown point of 50% and is labeled as LMedS. Recall from Section 3.1.1 that $\binom{M}{3}$ subgroups are built to determine the LS estimate, meaning the authors in [18] assume only one measurement per FT. Since we have K measurements per FT, we construct $K\binom{M}{3}$ subgroups and calculate the LS position estimates and corresponding residuals from each subgroup.

Then, the position estimate corresponding to the minimum of the *medians* taken over the squared residuals is the final position estimate. As stated in Section 3.1.1, the performance of this method depends on the optimization algorithm used to determine the LS estimate. For calculating the $K\binom{M}{3}$ LS position estimates we choose the Gauss-Newton algorithm [11] which compares to the Newton-Raphson algorithm in terms of computational complexity and thus maintains the overall computational power on a reasonable level. Remember that the latter is used for the semi-parametric approach. We use 10,000 Monte-Carlo runs and evaluate the mean error distance (MED), also known as circular positioning error [41], defined as

$$\text{MED} = \frac{1}{10^4} \sum_{i=1}^{10^4} \underbrace{\sqrt{(\hat{x}^{(i)} - x^{(i)})^2 + (\hat{y}^{(i)} - y^{(i)})^2}}_{\triangleq \text{location error}}, \qquad (3.23)$$

where $\hat{x}^{(i)}$ and $\hat{y}^{(i)}$ are the x- and y- component of the position estimate, and $x^{(i)}$ and $y^{(i)}$ are the true positions for the i-th Monte-Carlo run, respectively. Another performance metric is the RMSE defined as

$$\text{RMSE} = \sqrt{\frac{1}{10^4} \sum_{i=1}^{10^4} (\hat{x}^{(i)} - x^{(i)})^2 + (\hat{y}^{(i)} - y^{(i)})^2}. \qquad (3.24)$$

The sensor noise pdf is zero-mean Gaussian with standard deviation $\sigma_G = 150$m [111]. For the pdf f_η, modeling the NLOS effects, a shifted Gaussian and an exponential are used because they are widespread in the literature [22, 67, 92]. The parameters of these distributions are chosen according to typical values encountered in practical applications [92, 111]. Since all estimators become biased as ε increases, the CRLB is not meaningful anymore and thus not computed here. The different estimators and the parameters they require are summarized in Table 3.1.

Estimator	Required Quantity	Parameters	Reference
LS	-	-	Section 3.2.1
WLS	σ_G^2	-	Section 3.2.1, [23]
R-MR	-	$c_1 = 1.5, c_2 = 2.5$	Section 3.2.2
SP-MR	-	λ is set by (3.22), $\lambda < 1$, $\hat{\delta} = 1.06\hat{\sigma}_W(\dim(\mathbf{w}))^{-1/5}$	Section 3.2.3
LMedS	σ_G^2	-	Section 3.1.1, [18]

Table 3.1: Configurations of the estimators used throughout the simulations.

3.3.2 Simulation Results

3.3.2.1 Impact of λ on the Position Estimates of the Semi-Parametric Estimator

First, we want to assess the impact of λ for the SP-MR estimator on the RMSE of the position estimate $\hat{\boldsymbol{\theta}}$. To do so, in Step 3) of the algorithm explained in Figure 3.3, λ is not estimated but set to a certain value, $\hat{\boldsymbol{\theta}}$ is determined and the RMSE in (3.24) is evaluated. Recall that λ is the shape parameter of the transformation function $t(v, \lambda)$ in (3.21), depicted in Figure 3.2.

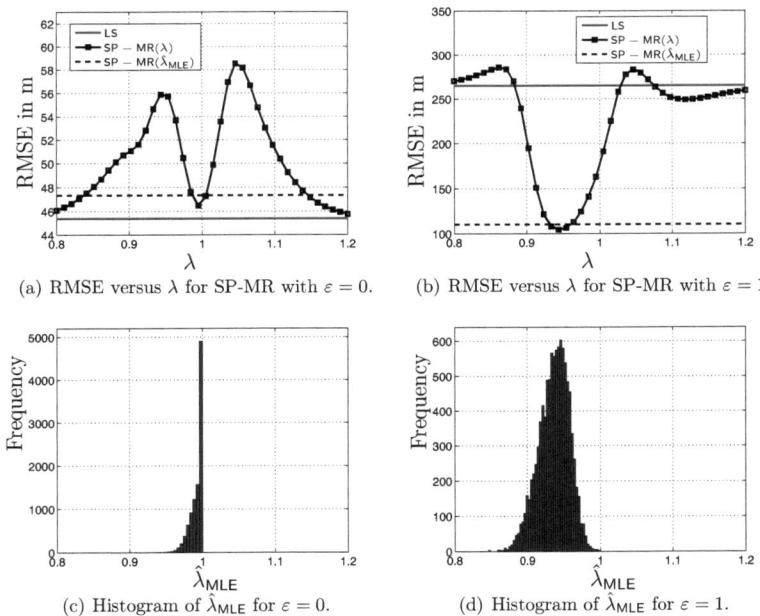

(a) RMSE versus λ for SP-MR with $\varepsilon = 0$.
(b) RMSE versus λ for SP-MR with $\varepsilon = 1$.
(c) Histogram of $\hat{\lambda}_{\text{MLE}}$ for $\varepsilon = 0$.
(d) Histogram of $\hat{\lambda}_{\text{MLE}}$ for $\varepsilon = 1$.

Figure 3.5: Figures (a) and (b) illustrate the RMSE for the different estimators when $\varepsilon = 0$ and $\varepsilon = 1$, respectively. For the SP-MR(λ), λ is fixed, whereas the estimator SP-MR($\hat{\lambda}_{\text{MLE}}$) chooses λ according to the MLE in (3.22). Figures (c) and (d) illustrate the histograms of $\hat{\lambda}_{\text{MLE}}$ for $\varepsilon = 0$ and $\varepsilon = 1$. For $\varepsilon = 1$, f_η is an exponential pdf with $\sigma_\eta = 500$m.

Figures 3.5(a) and 3.5(b) illustrate the RMSE of $\hat{\boldsymbol{\theta}}$ computed by SP-MR estimator in terms of the preset parameter λ, denoted as SP-MR(λ). For comparison purposes the RMSEs of two different estimators which do not depend on the preset λ are also depicted:

1. The LS estimator for $\hat{\boldsymbol{\theta}}$ is given as a reference because it is used to compute the starting point for the SP-MR algorithm.

2. The SP-MR estimator where λ is calculated by the MLE in (3.22) (with constraint $\lambda < 1$) and consequently does not depend on λ in the x-axis. This estimator is labeled as SP-MR($\hat{\lambda}_{\text{MLE}}$) in Figure 3.5.

We study the behavior in an environment where 0% (Figure 3.5(a)) and then 100% (Figure 3.5(b)) of the measurements are contaminated by NLOS errors, and f_η is the exponential pdf with $\sigma_\eta = 500$m. The interference is considered to be iid and $K = 5$ measurements are captured by each FT.

We observe in Figure 3.5(a) that the accuracy of the SP-MR(λ) strongly depends on the choice of λ. In LOS, $f_{\tilde{V}}(\tilde{v})$ is almost symmetric. Thus, the value of λ that minimizes the RMSE is close to one, meaning transformation of the residuals is not necessary, refer to Figure 3.2 for $\lambda = 1$. It is interesting to note that the MLE in (3.22) in most cases chooses λ close to one which is depicted in Figure 3.5(c). This consequently leads to high accuracy for the SP-MR($\hat{\lambda}_{\text{MLE}}$) estimator and its RMSE is similar to the RMSE of SP-MR($\lambda \approx 1$), which yields a local minimum. Note that the RMSE of SP-MR(λ) increases when λ slightly differs from one. This is because the symmetric, original residuals are transformed into an asymmetric sample that is symmetrized, i.e., $\mathbf{w}_s = [+t(\hat{\tilde{\mathbf{v}}}, \lambda) \ -t(\hat{\tilde{\mathbf{v}}}, \lambda)]$ leading to a model mismatch of the pdf estimate which increases the RMSE. Contrary, when λ further deviates from one, the RMSE of SP-MR(λ) is reduced again. In this case the pdf estimate becomes strongly biased because the sample is spread further apart. Then, the Newton-Raphson algorithm does not converge anymore and the RMSE of the semi-parametric estimate is similar to the initial LS estimate.

In contrast, if $\varepsilon = 1$, $f_{\tilde{V}}(\tilde{v})$ and consequently the residuals get right-skewed and have to be transformed such that their right tail is transformed closer to the core of the data. This can be achieved when $\lambda < 1$, refer to Figure 3.2. Numerical results in Figure 3.5(b) confirm this argument and the minimum of the RMSE of SP-MR(λ) is close to $\lambda = 0.95$. We can also verify in Figure 3.5(d) that the MLE for λ most often chooses values of λ around 0.95. Thus, high accuracy of the SP-MR($\hat{\lambda}_{\text{MLE}}$) estimator is achieved. Its RMSE is close to the global minimum RMSE of the SP-MR(λ) estimator. Again, when the value of λ is chosen farther away from the minimum, performance decreases due to model mismatch. Finally the RMSE for λ far away from the minimum is similar to the LS estimator since it is used for initialization of the SP-MR(λ) estimator which does not converge if the pdf is not estimated accurate enough.

Similar results are obtained for different values of ε and different NLOS error pdfs.

We choose the upper bound of λ in the left neighborhood of one to ensure that the right tail is always compressed, cf. Section 3.2.3.3, which is consistent with the results obtained in Figure 3.5. However, even though in general the MLE for λ (3.22) in Section 3.2.3.3 provides appropriate values for λ close to one, we set the lower bound of λ arbitrarily to 0.1 to ensure that the MLE does not drift too far away from the stable region which also reduces computational load for the selection of λ. In the sequel the semi-parametric estimator that automatically selects λ according to (3.22) for determining the pdf and the position is labeled as SP-MR.

3.3.2.2 NLOS Outliers Modeled as a Shifted Gaussian pdf

First, we investigate the performance of the different estimators in terms of MED versus the probability of NLOS occurrence ε. We consider a shifted Gaussian pdf with mean $\mu_\eta = 1000$m and standard deviation $\sigma_\eta = 300$m for modeling the NLOS errors where the moments are chosen according to typical values encountered in practical applications [92, 111]. Results are depicted in Figure 3.6.

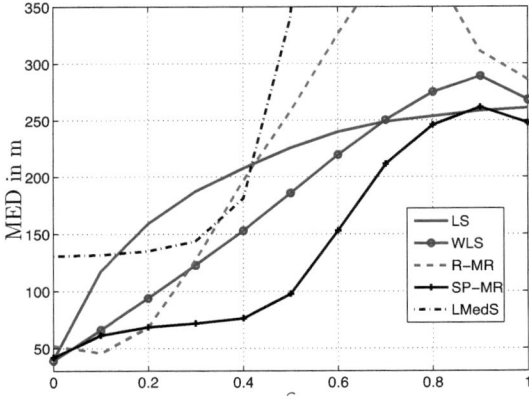

Figure 3.6: MED versus ε where a shifted Gaussian pdf with $\mu_\eta = 1000$m and $\sigma_\eta = 300$m are used for modeling NLOS errors. $K = 5$ measurements per FT are used for positioning.

It can be observed that for $\varepsilon = 0$ similar performance for all estimators is observed, except for LMedS which loses significantly in positioning accuracy. This is because it based on the *median* that provides a small efficiency under the Gaussian model [51]. However, when ε increases LS and WLS breaks down. Note that WLS achieves higher accuracy than LS since it underweighs observations with large noise samples. In contrast, R-MR achieves highest accuracy for small ε and breaks down when ε exceeds 30%. The reasons for this are the inability to adapt to the data and the model mismatch of the nominal Gaussian pdf within a certain region around zero. The LMedS only loses minor positioning accuracy up to $\varepsilon = 40\%$. This is expected and consistent with the theory [51] since the *median* has a breakdown point of 50%. At that percentage, the estimator is supposed to break down which is consistent with Figure 3.6. SP-MR significantly outperforms all its competitors for $0.3 < \varepsilon < 0.7$ whereas the highest gain in positioning accuracy is almost 90m at $\varepsilon = 0.5$. The superior performance of SP-MR comes from the fact that it is able to approximate the true underlying noise distribution.

Now we investigate the behavior of all estimators more thoroughly in the LOS case. The cdf of the MED for $\varepsilon = 0$ is depicted in Figure 3.7.

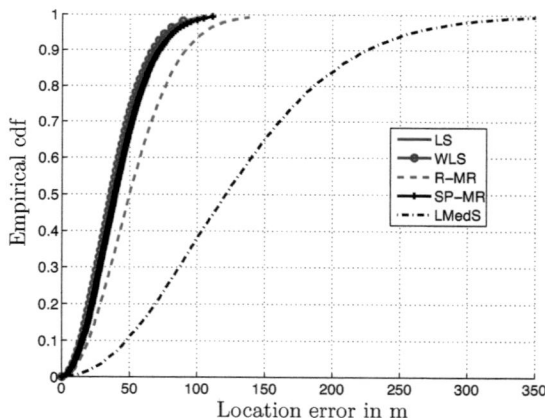

Figure 3.7: Empirical cdf of location errors in LOS environments ($\varepsilon = 0$) where $K = 5$ measurements per FT are used for positioning.

We observe that the estimator WLS achieves best performance in the LOS case followed by LS which is expected. The SP-MR loses up to 2m positioning accuracy with respect to WLS because it needs to estimate the underlying pdf and consequently has more variability than the parametric LS approaches. For the 95-percentile R-MR loses up to 15m positioning accuracy with respect to WLS because the redescending score function discards useful information by clipping the observations. The estimator LMedS [18] has lowest performance due to the efficiency loss of the *median* in the Gaussian case and achieves the 95-percentile at an MED of 250m.

To study the confidence bounds of the different estimators in NLOS environments we plot the cdf of the MED for $\varepsilon = 0.4$ illustrated in Figure 3.8. We observe that SP-MR significantly outperforms its competitors and has a 95-percentile of approximately 150m whereas WLS and LMedS have 95-percentiles at about 300m and LS and R-MR at 400m, respectively. The performance loss of LS and WLS is obvious due to linear weighting of the observations and consequently outliers. In contrast, R-MR has problems of convergence due to clipping of useful data. LMedS suffers from convergence problems in the Gauss-Newton algorithm used for determining the LS position estimates, caused by outliers. These problems of convergence result in outlying position estimates which become apparent in the upper percentiles of the cdf plot.

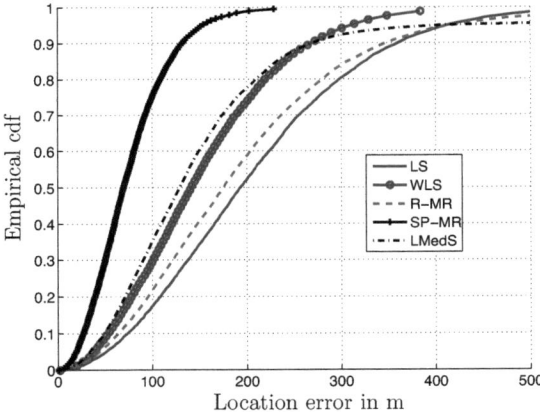

Figure 3.8: Empirical cdf of location errors in NLOS environments with $\varepsilon = 0.4$ and a shifted Gaussian pdf with $\mu_\eta = 1000$m and $\sigma_\eta = 300$m is used for modeling NLOS errors. $K = 5$ measurements per FT are used for positioning.

3.3.2.3 NLOS Outliers Modeled as an Exponential pdf

Since it is questionable whether using a shifted Gaussian pdf for modeling the NLOS error is realistic, we now study the performance of the different estimators in another scenario where f_η is an exponential pdf with $\sigma_\eta = 500$m [22, 111]. We vary ε and calculate the MED. The results are depicted in Figure 3.9 and it can be observed that the LS and WLS are outperformed by R-MR and SP-MR when $\varepsilon > 0.1$ and $\varepsilon > 0.2$, respectively. The R-MR improves positioning accuracy of about 20m and SP-MR gains approximately 40m in positioning accuracy with respect to WLS when $\varepsilon = 1$. Note that the gain in performance of SP-MR is more pronounced as in the previous example from Section 3.3.2.2. Furthermore, the positioning errors of all methods, except for the LMedS, are decreased due to smaller magnitudes of the NLOS outliers. This highlights the relative performance loss of LMedS. Note that accuracy of this estimator can be increased when using more complex optimization tools for selecting the LS estimates from the $K\binom{M}{3}$ subgroups.

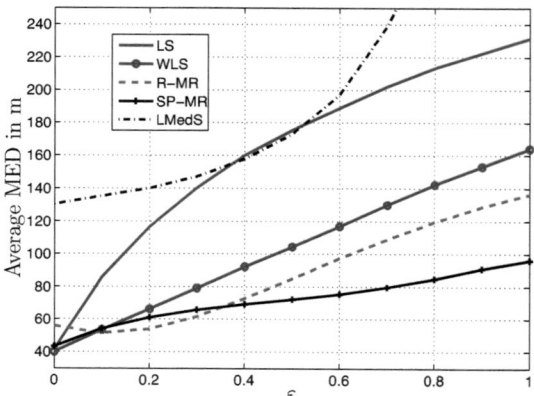

Figure 3.9: Average MED versus ε where f_η is an exponential pdf with $\sigma_\eta = 500$m and $K = 5$ measurements per FT are used for positioning.

Influence of NLOS Error Variance and Sample Size

Since we can not rely on $\sigma_\eta = 500$m to be realistic in any situation, the question arises about what happens if the standard deviation of the pdf f_η increases. For this purpose, another simulation study is conducted where $\varepsilon = 0.4$ of the observations are contaminated by NLOS outliers and σ_η is increased from 100m to 2500m. Results are illustrated in Figure 3.10 and it can be observed that LS and R-MR break down as soon as σ_η exceeds 500m and 1300m, respectively. The WLS estimator loses accuracy with respect to R-MR for small σ_η but gains in performance when σ_η exceeds 1100m. This is due to the fact that R-MR clips useful data, leading to divergence of the algorithm in Figure 3.1. It is interesting to note that LMedS remains stable for any value of σ_η and even improves slightly when σ_η increases. The reason for this behavior lies in the fact that the LMedS is based on ordering the squared residuals from different LS estimates corresponding to different subgroups. If the NLOS outliers in a subgroup become more pronounced, the residuals squared differ more severely from the residuals calculated from a subgroup with no NLOS outlier. This simplifies the identification of LS estimates from subgroups containing NLOS outliers resulting in smaller positioning errors by overweighing the accurate LS estimates. Furthermore, we observe that SP-MR significantly outperforms its competitors for $\sigma_\eta > 700$m and achieves similar positioning accuracy for smaller σ_η. However, in contrast to LMedS the MED of SP-MR slightly increases when σ_η grows. This is because the larger σ_η, the more difficult it becomes to estimate the pdf $f_{\tilde{V}}(\tilde{v})$ properly since $f_{\tilde{V}}(\tilde{v})$ spans over a larger region while the total number of observations does not change. If σ_η increases further, which is not necessarily realistic, it is expected that SP-MR loses slightly in performance whereas LMedS remains stable. Thus, at some σ_η LMedS can outperform SP-MR.

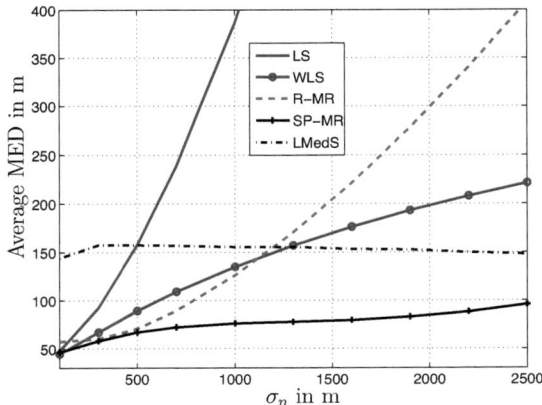

Figure 3.10: Average MED versus magnitude of NLOS standard deviation σ_η where f_η is an exponential pdf and $\varepsilon = 0.4$ of the measurements are contaminated by NLOS errors. $K = 5$ measurements per FT are available.

However, up to now we have considered $K = 5$ measurements per FT which gives us $MK = 50$ observations for positioning. The question that arises is how do the estimators, in particular SP-MR, perform in environments with a small number of measurements. To investigate this, we conduct a simulation study where K is changed from 1 to 10 and the results are shown in Figure 3.11 for $\varepsilon = 0.4$. We observe that all estimators gain in performance when K increases. However, even for $K = 1$, SP-MR has highest accuracy followed by WLS which loses about 15m and R-MR and LMedS which lose approximately 30m. The LS estimator completely breaks down. In contrast to SP-MR and WLS, R-MR rejects a large amount of observations and consequently, data for estimation, which leads to the performance loss.

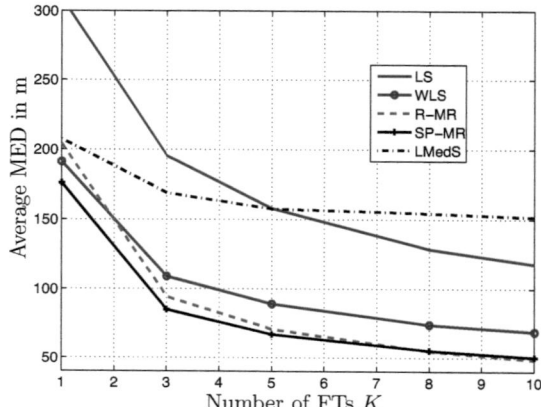

Figure 3.11: MED versus number of measurements per FT K where f_η is an exponential pdf with $\sigma_\eta = 500$m and $\varepsilon = 0.4$ of the measurements are contaminated by NLOS.

For larger K, R-MR achieves similar accuracy to SP-MR. This is because R-MR possesses over an appropriate amount of data so that even after discarding a large amount of measurements enough observations remain to achieve accurate estimates. The LMedS does only slightly gain in performance with increasing K which is again due to convergence problems of the Gauss-Newton algorithm caused by outliers or bad initialization. LS gains for $K > 5$ with respect to LMedS since enough observations are available to average out the effect of outliers.

For larger ε, the gain in accuracy of SP-MR becomes more significant. However, for $\varepsilon = 0$ and $K = 1$, SP-MR loses up to 10m in positioning accuracy with respect to LS and WLS since it has more variability than its parametric counterparts.

Note that, a similar effect can be observed when the number of FTs M is increased. Although not shown here, different models for f_η such as the uniform [18], the Rayleigh [53] and lognormal pdf were tested and similar results were obtained.

3.3.2.4 Comments on Computational Complexity

The LS estimator is the least complex algorithm in terms of computational power. It only requires to calculate the product of the linearized observation sequence and the pseudoinverse which can be computed offline. The WLS has a similar structure but is more complex because it requires computing the weights for the weighting matrix, and this has to be performed online.

Computation of the R-MR algorithm requires again the pseudoinverse, computed offline, as well as calculation of both the residuals and their scale estimate so that the normalized residuals fit the bounded score function. These steps are done in an iterative way and usually less than ten iterations are required for the iterative *modified residuals* algorithm to converge. In contrast the semi-parametric estimator SP-MR is more complex and apart from scale estimation it requires determination of λ, transformation of the residuals and KDE estimation. Furthermore, more iterations are required until convergence is achieved since parameter and pdf estimation are performed in parallel. On average, in the simulations presented within, it takes between 15 to 20 iterations for the SP-MR algorithm to converge.

The most complex algorithm is the LMedS because it requires to compute $K \binom{M}{3}$ nonlinear least-squares estimates. Each of them takes several iterations until convergence is achieved. Furthermore, the *median* from the squared residuals for each position estimates is computed $K \binom{M}{3}$ times.

3.4 Discussion and Conclusions

In this chapter, we have investigated the problem of finding the position of a stationary UE in NLOS environments using range measurements obtained from TOA estimates. Outliers due to NLOS propagation are modeled as positive random variables with various distributions, leading to a positively biased overall interference pdf which is not modeled by conventional techniques such as LS, WLS or robust M-estimators. These approaches break down or have limited positioning accuracy. Instead a semi-parametric estimator that estimates the interference pdf non-parametrically using TKDE is proposed. Based on the pdf estimate the position is obtained by using the ML principle. The degree of asymmetry, which changes between LOS and NLOS environments, is incorporated into the TKDE estimator to improve estimation accuracy.

Simulation results show that SP-MR significantly outperforms its competitors in NLOS environments while similar performance to the standard techniques is achieved in a LOS scenario. For the M-estimator, in the Gaussian case, an efficiency loss with respect to LS, WLS and SP-MR is observed whereas only in some cases a significant gain with respect to LS and WLS is achieved. This is because the redescending score functions can only slightly adapt to the noise by tuning their clipping parameters which is true for any parametric score function. Moreover, the assumption of symmetry in a certain region around zero for the noise pdf, required for the redescending score function, is only approximately fulfilled. In contrast, the estimator LMedS loses much precision in the Gaussian case but behaves robustly until $\varepsilon = 0.4$ when the noise power increases. However, although not shown here, the performance of this estimator strongly depends on the optimization algorithm used for determining the LS estimates for the different subgroups.

If the variability among $h_m(\boldsymbol{\theta})$ strongly increases, the errors in $f_{\tilde{V}}(\tilde{v})$ do not have the same variance anymore and $\tilde{\mathbf{v}}$ becomes heteroscedastic, which makes all estimators lose in positioning accuracy. In this case, it seems that WLS is least affected since the different variances are approximated by the covariance matrix. However, since WLS does not incorporate the prior knowledge of asymmetry, performance can be improved upon by taking this property into account. This can be achieved by using an algorithm that copes with asymmetric and heteroscedastic errors [109].

For different signal features such as AOA, TDOA and RSS, linearization of the measurement model can lead to stochastic elements and consequently to outliers in the regressor matrix, also called leverage points. These outliers lead to a performance loss in terms of positioning accuracy when using M-estimators or semi-parametric estimators. Improvements can be obtained by using generalized M-estimators [68] that downweigh outliers in the regressor matrix and in the observations and hence provide robust position estimates. SP-MR can be extended to these cases by incorporating a weighting function so that errors in the regressor matrix lose their deleterious impact on the parameter estimate.

Chapter 4
Robust Tracking

In this chapter, the problem of tracking a moving UE in mixed LOS/NLOS environments using TOA range measurements is treated. Applications arise in tracking fire-fighters and miners, locating lost children, intruder detection, home automation and patient monitoring, and intelligent transport systems among others [35].
Unlike in the previous chapter, where a set of measurements is recorded and processed at once (batch-processing), here, at each time step k, new measurements become available and are used to update the previous position estimate based on some motion model. For this reason, recursive Bayesian algorithms introduced in Section 2.2 become appealing. Since optimal approaches are intractable, suboptimal techniques such as the EKF or UKF are preferred. Even though theses standard techniques achieve high precision in LOS environments (Gaussian sensor noise only), they break down when large outliers due to NLOS propagation occur in the observations. Therefore, robust tracking algorithms are required.
Here, we adapt the semi-parametric estimator, presented in the previous chapter, to the tracking problem in a Kalman filter framework to provide a noise-adaptive tracker [45] that copes with LOS and NLOS environments. Furthermore, a multiple model algorithm is developed that uses a conventional EKF in parallel with a robustified EKF based on robust parametric M-estimation [47]. Depending on the situation one of them is overweighed, allowing for high accuracy in both LOS and NLOS environments. It is shown that the proposed trackers [45, 47] significantly outperform classical and robust competing trackers in different NLOS scenarios.

4.1 Problem Statement

4.1.1 Signal Model

Consider a moving UE with state vector $\mathbf{x}(k) = [x(k)\ y(k)\ \dot{x}(k)\ \dot{y}(k)]^\mathsf{T}$, as in Section 2.2.1, where $\dim(\mathbf{x}(k)) = N_x = 4$ here. The UE, surrounded by M FTs, is moving on a 2D-plane and its movement is described by the change of $\mathbf{x}(k)$ according to [41]

$$\mathbf{x}(k) = \mathbf{A}\mathbf{x}(k-1) + \mathbf{G}\boldsymbol{\omega}(k-1), \tag{4.1}$$

with

$$\mathbf{A} = \begin{bmatrix} \mathbf{I}_2 & \Delta t \cdot \mathbf{I}_2 \\ 0 & \mathbf{I}_2 \end{bmatrix}, \quad \mathbf{G} = \begin{bmatrix} \Delta t^2/2 \cdot \mathbf{I}_2 \\ \Delta t \cdot \mathbf{I}_2 \end{bmatrix}, \quad (4.2)$$

where Δt is the sampling period and $\mathbf{I}_M$ is the $M \times M$ identity matrix. The vector-valued driving noise $\boldsymbol{\omega}(k) = [\omega_1(k)\ \omega_2(k)]^\mathsf{T}$ is assumed zero-mean, white Gaussian with covariance matrix $\mathbf{Q}(k)$, $k = 1, 2, \ldots, K$, describing the uncertainty on the motion model at time k. The matrix $\mathbf{G}$ describes the mapping of the random accelerations contained in $\boldsymbol{\omega}(k)$ to the position and velocity of the UE and $\mathbf{A}$ is the state transition matrix describing the movement of the UE between two consecutive time steps. Let $\mathbf{y}(k)$ denote the vector of TOA estimates from M FTs, multiplied by the speed of light. Then,

$$\mathbf{y}(k) = \mathbf{h}(\mathbf{x}(k)) + \mathbf{v}_l(k), \quad k = 1, 2, \ldots, K, \ l = 1, 2, \quad (4.3)$$

where $\mathbf{h}(\mathbf{x}(k)) = [h_1(\mathbf{x}(k)), h_2(\mathbf{x}(k)), \ldots, h_M(\mathbf{x}(k))]^\mathsf{T}$ with,

$$h_m(\mathbf{x}(k)) = \sqrt{(x(k) - x_{\mathsf{FT},m})^2 + (y(k) - y_{\mathsf{FT},m})^2}, \quad m = 1, 2, \ldots, M \quad (4.4)$$

describes the true Euclidean distances between the UE and the m-th FT at time k. The noise vector $\mathbf{v}_l(k)$, $l = 1, 2$ describes Gaussian sensor noise and perturbations due to NLOS propagation. For this vector, we consider two different models, the nonlinear system model and the jump-nonlinear model described in Section 2.2.1.1 and Section 2.2.1.2, respectively, i.e,

$$\mathbf{v}_l(k) = \begin{cases} \mathbf{v}(k) & = [v_1(k), \ldots, v_M(k)]^\mathsf{T}, & l = 1, \\ \mathbf{v}(k, \mathcal{M}(k)) & = [v_1(k, \mathcal{M}(k)), \ldots, v_M(k, \mathcal{M}(k))]^\mathsf{T}, & l = 2, \end{cases} \quad (4.5)$$

For $l = 1$, $\mathbf{v}(k)$ is independently distributed over time and FTs (with $f_V(v) = (1 - \varepsilon)\mathcal{N}(v; 0, \sigma_G^2) + \varepsilon \mathcal{H}(v))$. In contrast, for the jump-nonlinear model, i.e., $l = 2$, $v_m(k, \mathcal{M}(k))\ \forall m = 1, 2, \ldots, M$ changes according to a two-state Markov chain (MC), depicted in Figure 2.5, modeling LOS/NLOS transitions. In LOS we have Gaussian sensor noise with variance σ_G^2 whereas in NLOS environments random quantities are drawn from a distribution with positive mean, i.e., $\mathcal{H}(v) = \mathcal{N}(v; 0, \sigma_G^2) * f_\eta$. For both models, the measurement covariance matrix $\mathbf{R}(k)$ is defined as

$$\mathbf{R}(k) = \mathsf{diag}[\sigma_1^2, \sigma_2^2, \ldots, \sigma_M^2], \quad (4.6)$$

where the elements σ_m^2 for the nonlinear system model and the jump-nonlinear model are defined in Table 4.1. Note that the probability of NLOS occurrence ε in the jump-nonlinear model can be calculated by (2.36).

4.1 Problem Statement

noise vector	diagonal elements of covariance matrix $\mathbf{R}(k)$ and pdf	l
$\mathbf{v}(k)$	$\sigma_m^2 = (1-\varepsilon)\sigma_G^2 + \varepsilon\sigma_\eta^2$ with $f_V(v) = (1-\varepsilon)\mathcal{N}(v;0,\sigma_G^2) + \varepsilon\mathcal{H}(v)$	1
$\mathbf{v}(k,\mathcal{M}(k))$	$\sigma_m^2 = \begin{cases} \sigma_G^2 & \text{if } m\text{-th FT is in LOS with } f_V(v) = \mathcal{N}(v;0,\sigma_G^2) \\ \sigma_G^2 + \sigma_\eta^2 & \text{if } m\text{-th FT is in NLOS with } f_V(v) = \mathcal{H}(v). \end{cases}$	2

Table 4.1: Parameters for the nonlinear system model in Section 2.2.1.1 and the jump-nonlinear model in Section 2.2.1.2.

Assumptions:

- Sensor noise variance σ_G^2 is known
- Process noise covariance $\mathbf{Q}(k)$ is known
- Degree of NLOS occurrence ε and NLOS error statistics f_η with variance σ_η^2 and consequently true measurement noise covariance matrix $\mathbf{R}(k)$ are unknown. Instead, $\mathbf{R}^*(k)$ denotes the measurement noise covariance matrix set by the trackers.

4.1.2 State of the Art

Depending on the model, different tracking schemes are available. Apart from the fact that optimal approaches are computational demanding for nonlinear and non-Gaussian problems, cf. Section 2.2, they are not feasible here since f_η and ε are unknown.
In this section, suboptimal classical and robust state estimation algorithms for equations (4.1) and (4.3) are presented. Suboptimal approaches, well suited for a white noise sequence $\mathbf{v}(k)$ ($l = 1$), include the EKF and a robustified version of it which are both introduced in the sequel. These techniques can be used in a multiple model framework to accommodate the switching of $\mathbf{R}(k)$ in the jump-nonlinear model (2.39) with noise sequence $\mathbf{v}(k,\mathcal{M}(k))$. The concepts of multiple model filters are introduced in Section 4.1.2.3 and a summary of other approaches found in the literature for tracking a UE is provided in Section 4.1.2.4.

4.1.2.1 Extended Kalman Filter

The most common approaches for the tracking problem (4.1) and (4.3), with white noise sequence $\mathbf{v}(k)$, are the KF for linear state and measurement equations and the EKF if one or both of the equations are nonlinear [5]. Both consist in calculating the conditional mean $\mathsf{E}\{\mathbf{x}(k)|\mathbf{Y}^k\}$ and associated covariance under the assumption that $\mathbf{x}(k)$, $k = 1, 2, \ldots, K$ is Gaussian. Thus, recursively calculating these moments completely describes the Gaussian pdf.

To sketch a derivation of the EKF we recall equations (2.40)-(2.43) from Section 2.2. Assume that $\mathbf{v}(k)$ and $\boldsymbol{\omega}(k)$ are white Gaussian (which is true for $\varepsilon = 0$), mutually independent processes independent from the initial state estimate $\hat{\mathbf{x}}(k-1|k-1) \sim \mathcal{N}(\mathbf{x}(k-1), \mathbf{P}(k-1|k-1))$ which can be obtained, e.g., by LS estimation. Then, $f(\mathbf{x}(k)|\mathbf{x}(k-1))$ and $f(\mathbf{x}(k-1)|\mathbf{Y}^{k-1})$ in (2.40) are Gaussian and we take the expectation $\mathsf{E}\{\mathbf{x}(k)|\mathbf{Y}^{k-1}\}$ and the corresponding covariance $\mathsf{Cov}\{\hat{\mathbf{x}}(k|k-1)\}$ for the prediction step yielding

$$\hat{\mathbf{x}}(k|k-1) = \mathbf{A}\hat{\mathbf{x}}(k-1|k-1) \quad (4.7)$$
$$\mathbf{P}(k|k-1) = \mathbf{A}\mathbf{P}(k-1|k-1)\mathbf{A}^\mathsf{T} + \mathbf{G}\mathbf{Q}(k)\mathbf{G}^\mathsf{T}, \quad (4.8)$$

which completely describe $f(\mathbf{x}(k)|\mathbf{Y}^{k-1})$. To obtain the posterior pdf (2.43), required to calculate the conditional mean estimate, it is necessary to transform the Gaussian rv $\hat{\mathbf{x}}(k|k-1)$ by the nonlinear function $\mathbf{h}(\cdot)$, leading to a non-Gaussian conditional pdf $f(\mathbf{x}(k)|\mathbf{Y}^k)$. Describing $f(\mathbf{x}(k)|\mathbf{Y}^k)$ by its first two moments is then insufficient because higher order moments occur for non-Gaussian data. Instead, to describe the entire statistics by only two moments, the posterior density $f(\mathbf{x}(k)|\mathbf{Y}^k)$ is approximated by a Gaussian pdf. To achieve this, we first linearize $\mathbf{h}(\cdot)$ around the predicted state estimate $\hat{\mathbf{x}}(k|k-1)$ by using a first-order Taylor approximation, i.e.,

$$\mathbf{h}(\mathbf{x}(k)) \approx \mathbf{h}(\hat{\mathbf{x}}(k|k-1)) + \underbrace{\frac{\partial \mathbf{h}(\mathbf{x}(k))}{\partial \mathbf{x}(k)}\bigg|_{\mathbf{x}(k)=\hat{\mathbf{x}}(k|k-1)}}_{=\mathbf{H}(k)}(\mathbf{x}(k) - \hat{\mathbf{x}}(k|k-1)) \quad (4.9)$$

where

$$\mathbf{H}(k) = \frac{\partial \mathbf{h}(\mathbf{x}(k))}{\partial \mathbf{x}(k)}\bigg|_{\mathbf{x}(k)=\hat{\mathbf{x}}(k|k-1)}$$

$$= \begin{bmatrix} \frac{dh_1(\mathbf{x}(k))}{dx(k)}\big|_{\mathbf{x}(k)=\hat{\mathbf{x}}(k|k-1)} & \frac{dh_1(\mathbf{x}(k))}{dy(k)}\big|_{\mathbf{x}(k)=\hat{\mathbf{x}}(k|k-1)} & 0 & 0 \\ \vdots & \vdots & \vdots & \vdots \\ \frac{dh_M(\mathbf{x}(k))}{dx(k)}\big|_{\mathbf{x}(k)=\hat{\mathbf{x}}(k|k-1)} & \frac{dh_M(\mathbf{x}(k))}{dy(k)}\big|_{\mathbf{x}(k)=\hat{\mathbf{x}}(k|k-1)} & 0 & 0 \end{bmatrix}. \quad (4.10)$$

is the Jacobian matrix of partial derivatives evaluated at the predicted state estimate $\hat{\mathbf{x}}(k|k-1)$. Even though a rv transformed by a nonlinear function does not conserve its distribution, it is assumed that $\mathbf{h}(\hat{\mathbf{x}}(k|k-1))$ is Gaussian. Then, calculating the

4.1 Problem Statement

posterior pdf in (2.43) to compute the conditional mean (2.44) yields the EKF equations

$$\boldsymbol{\nu}(k) = \mathbf{y}(k) - \mathbf{h}(\hat{\mathbf{x}}(k|k-1)) \tag{4.11}$$
$$\mathbf{S}(k) = \mathbf{H}(k)\mathbf{P}(k|k-1)\mathbf{H}^\mathsf{T}(k) + \mathbf{R}(k) \tag{4.12}$$
$$\mathbf{K}(k) = \mathbf{P}(k|k-1)\mathbf{H}^\mathsf{T}(k)\mathbf{S}^{-1}(k) \tag{4.13}$$
$$\hat{\mathbf{x}}(k|k) = \hat{\mathbf{x}}(k|k-1) + \mathbf{K}(k)\boldsymbol{\nu}(k) \tag{4.14}$$
$$\mathbf{P}(k|k) = (\mathbf{I}_4 - \mathbf{K}(k)\mathbf{H}(k))\mathbf{P}(k|k-1), \tag{4.15}$$

where (4.11) describes the innovation sequence with covariance (4.12) corresponding to $f(\mathbf{y}(k)|\mathbf{Y}^{k-1}) = \mathcal{N}(\boldsymbol{\nu}(k); \mathbf{0}, \mathbf{S}(k))$. The Kalman gain is calculated in (4.13) to yield the posterior mean (4.14) and associated covariance (4.15). If the linearization errors in (4.9) become too large, information about the measurement model is lost, which can lead to high positioning errors and even convergence problems. For smaller linearization errors, the approximation holds and accurate positioning can be achieved in LOS environments. However, when outliers due to NLOS propagation in the observations $\mathbf{y}(k)$ and consequently in the innovation sequence (4.11) occur, large positioning errors are expected because the innovations have an unbounded effect in (4.14) on the state estimate $\hat{\mathbf{x}}(k|k)$. Thus, robust alternatives are required.

4.1.2.2 Robust Extended Kalman Filter

A common approach to robust Kalman filtering is to rewrite the KF equations into a linear regression problem and apply robust techniques [51], explained in Section 2.1.3, to solve for the state vector $\mathbf{x}(k)$. The equivalence between the linear KF and the LS solution of a linear regression problem is shown in [9, 28] and extended to the EKF in [29]. This relationship is exploited to apply robust regression techniques for each time step k. Here, we present this approach and the robust EKF (REKF), applied to hybrid positioning in [80], serves for comparison purposes later on. Furthermore, it is the basis for applying semi-parametric techniques in the EKF framework and to be used in a multiple model framework.

First, the state and measurement equations (4.1) and (4.3) are rewritten into

$$\begin{bmatrix} \mathbf{I}_4 \\ \mathbf{H}(k) \end{bmatrix} \mathbf{x}(k) = \begin{bmatrix} \mathbf{A}\hat{\mathbf{x}}(k-1|k-1) \\ \mathbf{y}(k) - \mathbf{h}(\hat{\mathbf{x}}(k|k-1)) + \mathbf{H}(k)\hat{\mathbf{x}}(k|k-1) \end{bmatrix} + \mathbf{e}(k), \tag{4.16}$$

where

$$\mathbf{e}(k) = \begin{bmatrix} \mathbf{A}(\mathbf{x}(k-1) - \hat{\mathbf{x}}(k-1|k-1)) + \mathbf{G}\boldsymbol{\omega}(k-1), \\ -\mathbf{v}(k) \end{bmatrix} \tag{4.17}$$

with
$$E[\mathbf{e}(k)\mathbf{e}^\mathsf{T}(k)] = \begin{bmatrix} \mathbf{P}(k|k-1) & \mathbf{0} \\ \mathbf{0} & \mathbf{R}(k) \end{bmatrix} = \mathbf{C}(k)\mathbf{C}^\mathsf{T}(k), \qquad (4.18)$$
where $\mathbf{P}(k|k-1)$ is defined in (4.8) and $\mathbf{C}(k)$ in (4.18) is obtained, e.g., by using Cholesky decomposition [28, 39]. Then, multiplying (4.16) with $\mathbf{C}^{-1}(k)$ yields the linear regression model
$$\tilde{\mathbf{y}} = \mathbf{D}\boldsymbol{\theta} + \tilde{\mathbf{v}}, \qquad (4.19)$$
where
$$\tilde{\mathbf{y}} = \mathbf{C}^{-1}(k) \begin{bmatrix} \hat{\mathbf{x}}(k|k-1) \\ \mathbf{y}(k) - \mathbf{h}(\hat{\mathbf{x}}(k|k-1)) + \mathbf{H}(k)\hat{\mathbf{x}}(k|k-1) \end{bmatrix}, \qquad (4.20)$$
$$\mathbf{D} = \mathbf{C}^{-1}(k) \begin{bmatrix} \mathbf{I}_4 \\ \mathbf{H}(k) \end{bmatrix}, \quad \boldsymbol{\theta} = \mathbf{x}(k), \quad \tilde{\mathbf{v}} = -\mathbf{C}^{-1}(k)\mathbf{e}(k).$$

In LOS, i.e., if $\mathbf{v}(k)$ is zero-mean Gaussian, $\mathsf{E}\{\tilde{\mathbf{v}}\} = \mathbf{0}$ and the covariance matrix of $\tilde{\mathbf{v}}$ is $\mathsf{E}\{\tilde{\mathbf{v}}\tilde{\mathbf{v}}^\mathsf{T}\} = \mathbf{I}_{4+M}$. The LS solution of (4.19), $\hat{\mathbf{x}}(k|k) = \hat{\boldsymbol{\theta}} = (\mathbf{D}^\mathsf{T}\mathbf{D})^{-1}\mathbf{D}^\mathsf{T}\tilde{\mathbf{y}}$, given in (3.8), is equivalent to the state estimate $\hat{\mathbf{x}}(k|k)$ obtained in (4.14) from the EKF [28]. The covariance matrix of the state estimate is then
$$\mathbf{P}(k|k) = \mathsf{Cov}\{\hat{\mathbf{x}}(k|k)\} = (\mathbf{D}^\mathsf{T}\mathbf{D})^{-1}, \qquad (4.21)$$
and equals the posterior covariance (4.15) of the EKF. However, since LS estimation is sensitive to outliers, we solve (4.19) using robust regression techniques as these explained in Section 3.2.2. For this purpose matrix $\mathbf{D}$ from (4.20) is assumed deterministic and the algorithm given in Figure 3.1 is used. Then $\hat{\mathbf{x}}(k|k)$ is asymptotically (for large M) Gaussian. This approach serves as a comparison for the semi-parametric tracker proposed in the next section. The posterior covariance of the robust state estimate $\hat{\mathbf{x}}(k|k)$ can be either estimated using (2.13) in Section 2.1.3 or approximated using (4.15) or (4.21). The latter approach is preferred here because it has been found that it is numerically more stable [45]. As stated in Section 3.2.2, the performance of any estimator based on a bounded score function strongly depends on the fit of the function to the underlying noise model and tuning of one or several clipping points [24, 49, 51]. Again, since the NLOS errors follow a distribution with positive mean, the resulting pdf $f_{\tilde{V}}(\tilde{v})$ is asymmetric and we propose to use the redescending score function [24, 49] given in Equation (3.15). As in the previous chapter the choice of the clipping points always requires a compromise between efficiency in the Gaussian case and robustness in the non-Gaussian case. In Appendix A.2.1 a numerical study on the mean error distance (MED) versus the clipping parameters c_1 and c_2 can be found. The limitations of parametric M-estimators have been extensively explained in Section 2.1.3 and become apparent in Chapter 3. They also hold for the underlying tracking problem. Instead we consider a semi-parametric approach in the EKF framework to circumvent these issues.

4.1.2.3 Interacting Multiple Model Algorithm and Existing Approaches

Reconsider measurement equation (4.3) where $\mathbf{v}_2(k) = \mathbf{v}(k, \mathcal{M}(k))$. To describe the time dependencies due to shadowing in the LOS channel, switching between LOS/NLOS events occurs at random time steps modeled according to a two-state MC at each FT. When considering M FTs, the augmented MC with $r = 2^M$ states is calculated as in (2.38,) resulting in the transition probability matrix (TPM)

$$\mathbf{T} = \begin{pmatrix} p_{11} & p_{12} & \cdots & p_{1r} \\ p_{21} & p_{22} & \cdots & p_{2r} \\ \vdots & \vdots & \ddots & \vdots \\ p_{r1} & p_{r2} & \cdots & p_{rr} \end{pmatrix}, \tag{4.22}$$

where the transition probabilities are $p_{ij} \in \mathbb{R}$, $0 \leq p_{ij} \leq 1 \ \forall \ i,j = 1,2,\ldots,r$ and $\sum_j p_{ij} = 1 \ \forall \ i = 1,2,\ldots,r$. The mode variable $\mathcal{M}(k)$ at time step k is among the r modes, i.e., $\mathcal{M}(k) \in \{\mathcal{M}_j\}_{j=1}^r$. This results in a hybrid estimation problem because apart from the continuous-valued state vector $\mathbf{x}(k)$ we need to implicitly estimate a discrete mode variable $\mathcal{M}(k)$ (even though the second variable is not of any interest to the user). For this problem, optimal solutions are intractable for two reasons (cf. Section 2.2.3). First, the parameters of the NLOS error statistics and transition probabilities in (4.22) of the augmented MC are unknown. Second, even if these parameters were known, an optimal solution requires conditioning on all mode sequences. This is not feasible since the number of mode sequences grows exponentially with $(2^M)^k$.

Instead, suboptimal approaches like the generalized pseudo-Bayesian (GPB) or interacting multiple model (IMM) algorithm [5] take into account information from the previous time step to calculate the state estimates and their covariances. Here, we consider the IMM algorithm because it trades off performance versus complexity in an appropriate manner [71]. It is assumed that the posterior pdfs conditioned on each of the r modes are Gaussian. The functionality of the IMM algorithm, sketched in Figure 4.1, is briefly discussed hereafter. For a detailed derivation see [5].

Assume that the entire past through $k-1$ is summarized by r mode-conditioned estimates and covariances, i.e, $\hat{\mathbf{x}}_i(k-1|k-1)$, $\mathbf{P}_i(k-1|k-1) \ \forall i = 1,2,\ldots,r$. *Interaction* of these estimates with *mixing probabilities* $\mu_{i|j}(k-1|k-1)$ containing the elements p_{ij} from (4.22) yield $\hat{\mathbf{x}}_{0j}(k-1|k-1)$, $\mathbf{P}_{0j}(k-1|k-1) \ \forall j = 1,2,\ldots,r$. These quantities are used to initialize each of the r filters matched to model $\mathcal{M}_j$ (*Mode-matched filtering*). Then, each filter such as a KF for linear systems and an EKF for nonlinear systems, estimates the quantities $\hat{\mathbf{x}}_j(k|k)$ and $\mathbf{P}_j(k|k)$ together with the likelihood $\Lambda_j(k)$ of model $\mathcal{M}_j$ being the correct one. This likelihood is used to calculate the *mode probabilities* $\mu_j(k)$, and the state estimates from the r filters are combined with these probabilities to yield the final state estimate $\hat{\mathbf{x}}(k|k)$ and covariance estimate $\mathbf{P}(k|k)$.

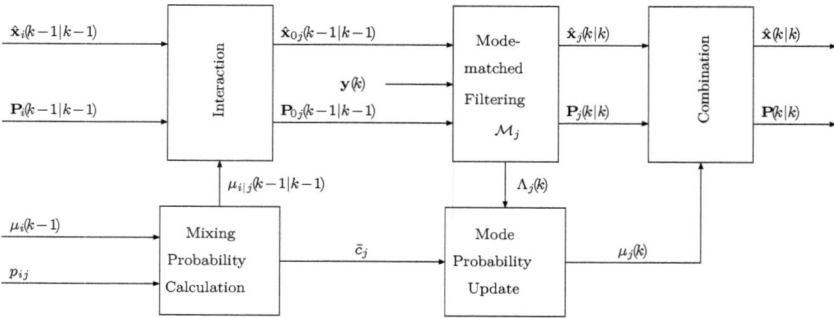

Figure 4.1: One cycle of the IMM algorithm

The IMM algorithm has been applied to the NLOS problem in [20, 33, 67]. In [67], an IMM algorithm with two KFs at each FT is used to smooth the TOA range estimates by distinguishing between the LOS/NLOS models. The state estimate in the NLOS mode, where the bias of the TOA range estimates is subtracted, is combined with the state estimate in LOS mode in order to determine the ranges between the UE and the FTs, irrespective which FT is in LOS or NLOS. The final position is obtained from the smoothed range estimates by using a geometric method. A similar approach for UE tracking based on RSS and TOA data fusion is suggested in [20].

The authors in [33] consider TOA measurements and suggest to use $r = 2^M$ EKFs matched to each mode $\mathcal{M}_j \; \forall j = 1, 2, \ldots, r$ in parallel, where the innovation sequence in each EKF is corrected for the bias, and the measurement covariance in each EKF is adapted to the covariance of the LOS/ NLOS error statistics. This approach achieves close to optimal performance in both LOS and NLOS environments. However, in practice, the parameters of the NLOS error statistics are unknown and even time-varying which make the approaches infeasible. Furthermore the complexity of the latter algorithm increases exponentially with the number of available FTs since we have to run $r = 2^M$ filters in parallel. For these reasons, it is desirable to design an algorithm that adapts to different LOS/NLOS situations (without having any knowledge of the NLOS error statistics) while keeping the number of filters used in parallel at a reasonable level. In this context, an IMM algorithm that uses REKF and EKF in parallel is proposed in Section 4.3.

4.1.2.4 Other Existing Techniques for Tracking an UE

In some environments, when the nonlinearity $\mathbf{h}(\cdot)$ increases further, improvements can be achieved by using the UKF [103], the cubature KF [2] or particle filter (PF) [3] which better approximate the moments of the rv transformed by a nonlinear function. However, if the measurement noise deviates from the Gaussian assumption, which is true in NLOS environments, the UKF and cubature Kalman filter can significantly lose in positioning accuracy.

Given that the measurement noise density is known, the PF is able to cope with non-Gaussian error statistics but it is computational demanding.

Various tracking schemes for positioning of a moving UE in NLOS environments are available in the literature [4, 21, 53, 54, 77, 78, 111]. The authors in [111] suggest to use a hypothesis test based on the standard deviation of the measurement errors to discriminate between LOS and NLOS observations. A more sophisticated NLOS bias detection scheme is suggested in [4] where an UKF is used to track the UE. If a bias is detected, the corresponding observation is discarded in the update step of the UKF. In [53] the UKF is modified to incorporate prior information on LOS/NLOS measurements into the update step which is in general not available.

Other approaches which take advantage of the measurement time-history are presented in [77, 78] where the NLOS bias is estimated jointly with the position and velocity of the UE. These approaches achieve improvements in terms of positioning accuracy with respect to conventional ones but require augmentation of the state space to $N_x + M$ instead of N_x, leading to a higher computational load. The authors in [54] propose to model the LOS/NLOS occurrences as a time-varying stochastic process where the probability for LOS/NLOS of each measurement is estimated and used in a UKF to estimate the position of the UE. These quantities are exploited in the next time step to estimate the LOS/NLOS probabilities and so forth.

However, the above mentioned algorithms either rely on a certain noise distribution or assume a specific model for the occurrence of NLOS observations. Instead, Chen proposed a tracking algorithm [21] which uses the NLOS mitigation algorithm in [22], described in Section 3.1.1, together with a KF for UE positioning. This approach neither imposes any distribution on the NLOS error statistics nor any model on the time dependencies of the NLOS errors and serves as a reference here for comparison purposes.

4.2 Noise-adaptive EKF using Semi-Parametric Estimation

4.2.1 General Idea

For the EKF and REKF, the first and second moments of the state vector $\hat{\mathbf{x}}(k|k)$ are propagated through the EKF iterations which are written into a linear regression problem in Section 4.1.2.2. Solving this problem with LS or robust techniques corresponding to EKF and REKF, respectively, still maintains (at least asymptotically, for large M) Gaussianity for the updated state vector $\hat{\mathbf{x}}(k|k)$ irrespective of LOS or NLOS situations. However, in NLOS environments, given that f_η is non-Gaussian, higher order moments which are not exploited by the above mentioned estimators occur in $\mathbf{v}_l(k)$, $l = 1, 2$. Improvements can be achieved by incorporating these moments into the estimator meaning by learning from the measurements. For this purpose, the interference pdf of $\tilde{\mathbf{v}}(k)$ in (4.19) is estimated non-parametrically based on TKDE as explained in Section 3.2.3.2. This pdf estimate contains moments up to infinite order and is used to calculate the state vector $\hat{\mathbf{x}}(k|k)$ based on the ML principle. This estimate is then propagated through the prediction equations (4.7) and (4.8) to the next time step $k + 1$.

The asymptotic variance of the semi-parametric estimator has not yet been established and (4.21) is used instead as an approximation. An alternative, to obtain more accurate estimates of the posterior covariance, is to use resampling techniques, that are more computational demanding [114],.

To the best of the authors' knowledge, applying semi-parametric estimation techniques to Kalman filtering (standard or extended) has not been suggested in the literature as yet. Given that a large number of FTs are available in LOS environments, it is expected that the noise-adaptive EKF only slightly loses performance with respect to the conventional EKF since it possesses of more degrees of freedom.

However, when ε increases, the assumption $\mathsf{E}\{\tilde{\mathbf{v}}\tilde{\mathbf{v}}^\mathsf{T}\} = \mathbf{I}_{4+M}$ (from (4.20)) does not hold anymore because $\mathbf{R}^*(k)$ selected by the noise-adaptive EKF does not correspond to the true one $\mathbf{R}(k)$, which leads to a mismatch when multiplying the quantities in (4.20) with $\mathbf{C}^{-1}(k)$. Estimating the noise pdf non-parametrically corrects for this mismatch and consequently leads to higher precision with respect to EKF or REKF. Note that the mismatch is larger when the jump-nonlinear model is considered because, depending on σ_η, the variances of the noise samples $v_m(k, \mathcal{M}(k))$ for one time step k can differ strongly. This complicates estimation of the noise pdf and can result in a bias. However, since our goal is not to estimate the pdf but the parameter of interest, i.e., the state vector, it is more important to obtain smooth pdf estimates, even though biased, so that the minimum of the negative log-likelihood function can be found with a gradient search (e.g. here, Newton-Raphson). Note that the sample size of residuals is artificially enlarged and doubled, i.e., assuming a symmetric transformed density (cf. Section 3.2.3.2), we have $2(M + N_x)$ residuals that can be used for KDE in the W-domain. For even small sample sizes, the Newton-Raphson step is thus performed using more than twice as many data points.

4.2.2 Algorithm

The algorithm is based on the EKF equations in linear regression form, given in (4.19). A summary of the algorithm is given in Table 4.2. After an initial estimate $\hat{\mathbf{x}}(k-1|k-1)$ and its covariance $\mathbf{P}(k-1|k-1)$ for $k = 1$ are obtained, e.g., via LS, we calculate the predicted state vector $\hat{\mathbf{x}}(k|k-1)$ and its covariance $\mathbf{P}(k|k-1)$ with (4.7) and (4.8), respectively. Then, the EKF equations are rewritten into a linear regression model (4.19) and LS estimation is performed to obtain an initial state estimate $\hat{\mathbf{x}}(k|k)$ for $k = 2$. This estimate is then used to compute the residuals and the algorithm for semi-parametric estimation, given in Section 3.2.3.4 is used to improve accuracy of the state estimate $\hat{\mathbf{x}}(k|k)$. The posterior covariance matrix is approximated by $(\mathbf{D}^\mathsf{T}\mathbf{D})^{-1}$. Again, we suggest using the semi-parametric algorithm based on the *modified residuals* since the semi-parametric algorithm based on *modified weights* is numerically less stable and suffers from a mismatch between the computed state estimate and the associated covariance matrix [45].

Remember that only the first and second moments of $\hat{\mathbf{x}}(k|k)$ are propagated through the EKF iterations. However, all moments of the noise pdf are estimated for a particular time step (using non-parametric pdf estimation) and exploited for state estimation. This improves positioning accuracy with respect to traditional robust EKFs.

	For each time step k do
Step 1)	Compute the predicted state vector $\hat{\mathbf{x}}(k|k-1)$ and its covariance $\mathbf{P}(k|k-1)$ with (4.7) and (4.8), respectively.
Step 2)	Compute the Cholesky factorization in (4.18) and $\tilde{\mathbf{y}}$ and $\mathbf{D}$ in (4.20) to obtain the linear regression problem (4.19).
Step 3)	Solve (4.19) using the semi-parametric estimation algorithm illustrated in Figure 3.3, in Section 3.2.3.
Step 4)	Approximate the posterior covariance in (4.15) with $\mathbf{P}(k|k) = (\mathbf{D}^\mathsf{T}\mathbf{D})^{-1}$.

Table 4.2: One cycle of the semi-parametric (noise-adaptive) tracking algorithm.

4.3 Robust Tracking based on M-Estimation and Interacting Multiple Model Algorithm

As we have seen in Chapters 2 and 3, M-estimators always trade off efficiency in the nominal case versus robustness in the NLOS case and one or more clipping points have to be set beforehand to regulate the degree of robustness/efficiency. Thus, it is not possible to achieve efficient estimates in the LOS case together with robust estimates in the NLOS case using the same clipping points. Here, we propose using an IMM algorithm which inherently combines the state estimates of an EKF together with the state estimates of an REKF to achieve both high accuracy in LOS, Gaussian case and robustness in the NLOS, non-Gaussian case.

4.3.1 Model Reduction

If we knew σ_η^2, even roughly, we could model all $r = 2^M$ constellations of LOS/NLOS occurrences in parallel through adaptation of the 2^M measurement covariance matrices $\mathbf{R}(k)$ given in Table 4.1 for $l = 2$. Then, depending on the model, r filters (either non-robust or robust to a certain degree) could be used in parallel to mitigate NLOS outliers in the observations and to obtain accurate position estimates when LOS FTs are predominant. However, this approach, in particular for large M, implies a high computational load and is therefore not advised.

An alternative is to reduce the model from 2^M states to a smaller number of states where several states from the complete model with transition probabilities in (4.22) are mapped into one state of the reduced model. Here, we distinguish between two different models ($\mathcal{M}_j$ $j = 1, 2$) to decrease computational load further. This evokes two different approaches for modeling: Our first suggestion is to use an EKF (model $\mathcal{M}_1$ in the complete model), well suited for LOS environments and an REKF, for all other cases ($\mathcal{M}_1, \mathcal{M}_2, \ldots, \mathcal{M}_r$ in the complete model). In this case the TPM from (4.22) reduces to a 2×2 matrix $\mathbf{T}'$. Note that the transition probabilities of this matrix are calculated in Appendix A.2.2 using Bayes' theorem and a summary of the modeling from 2^M to two states is illustrated in Table 4.3.

For the EKF, the measurement covariance matrix is adapted to the sensor noise covariance. However, the main problem with this approach is that it is unclear how to determine the covariance matrix for the REKF because $r - 1$ different models with different covariance matrices have to be transformed into a single one. Furthermore, incorporating the new transition probabilities contained in $\mathbf{T}'$ can sometimes even lead to wrong weighting of the different filters. In particular when the NLOS error realizations are rather small, the EKF can result in more accurate position estimates since it does not cut any information, unlike the REKF.

Reduced Model	Complete Model	Filter	$\mathbf{R}(k)$	Trans. prob. of $\mathbf{T}'$
$\mathcal{M}'_1$	$\mathcal{M}_1$	EKF	$\sigma_G^2 \mathbf{I}_M$	$p'_{12} = \sum_{j=2}^{r} p_{1j}$
$\mathcal{M}'_2$	$\mathcal{M}_2, \mathcal{M}_3, \ldots, \mathcal{M}_r$	REKF	?	$p'_{21} = \dfrac{\sum_{j=2}^{r} p_{1j} P\{\mathcal{M}_1\}}{\sum_{j=2}^{r} P\{\mathcal{M}_j\}}$

Table 4.3: Reduced model is obtained by considering all 2^M models.

For these reasons it becomes desirable to choose a different reduced model. We propose to model the two extreme cases where either all FTs are in LOS, modeled by an EKF, or all FTs are in NLOS, modeled by an REKF. For the former, the measurement covariance matrix is given as $\mathbf{R}(k) = \sigma_G^2 \mathbf{I}_M$ and for the latter it is defined as $\mathbf{R}(k) = (\sigma_G^2 + \sigma_\eta^2)\mathbf{I}_M$. All other models ($\mathcal{M}_2, \mathcal{M}_3, \ldots, \mathcal{M}_{r-1}$) are not explicitly defined here. However, if one of these models is in effect, the state estimates can be obtained by linear combinations of the state estimates associated to the two different models mentioned above. This approach is followed in the sequel and a summary of this submodeling is illustrated in Table 4.4. The transition probabilities for the reduced model with TPM $\mathbf{T}'$ are the complements of the probabilities of the MC of remaining in one of the extreme cases. In general σ_η^2 as well as the transition probabilities p_{ij} and consequently p'_{ij} are unknown. However, we will show later in the simulation results that these quantities are not required to achieve robustness in NLOS environments and a similar performance to the EKF in LOS environments.

Reduced Model	Complete Model	Filter	$\mathbf{R}(k)$	Trans. prob. of $\mathbf{T}'$
$\mathcal{M}'_1$	$\mathcal{M}_1$	EKF	$\sigma_G^2 \mathbf{I}_M$	$p'_{12} = 1 - p_{11}$
$\mathcal{M}'_2$	$\mathcal{M}_r$	REKF	$(\sigma_G^2 + \sigma_\eta^2)\mathbf{I}_M$	$p'_{21} = 1 - p_{rr}$

Table 4.4: Reduced model is obtained by modeling the two extreme cases where either all FTs are in LOS or all FTs are in NLOS.

4.3.2 Algorithm

In this section, a description of the robust IMM algorithm, using the reduced model sketched in Table 4.4, is provided. The flow chart from Figure 4.1 is precised in Table

4.5 yielding the R-IMM algorithm.

At Step 1), the prior probabilities $\mu_1(k-1)$, $\mu_2(k-1)$ and the prior state and covariance estimates $\hat{\mathbf{x}}(k-1|k-1)$ and $\mathbf{P}(k-1|k-1)$, respectively, are initialized. The transition probabilities p'_{ij} of $\mathbf{T}'$ are not known in practice but can be chosen based on some prior knowledge or set to $p_{ij} = 0.5$ for $i,j = 1,2$. Then, the mixing probabilities $\mu_{i|j}(k-1|k-1)$ are calculated in (4.23) where the normalization constant $\bar{c}_j$ is chosen in (4.24) such that $\sum_{j=1}^{2} \mu_{ij}(k-1|k-1) = 1$. Then, at Step 2), interaction of the previous state and covariance estimates is performed in (4.25) and (4.26), respectively to provide initial estimates for the two mode-matched filters in Step 3). Here, the prediction steps of the EKF and REKF are run in parallel. For the latter in (4.28) a higher measurement covariance matrix $\mathbf{R}_2^*(k)$ is chosen to cope with the higher variances of the NLOS errors. For $j = 1$, the EKF update steps are computed to obtain the state and covariance estimates. In contrast, for $j = 2$, the EKF equations of the robust filter are rewritten into a linear regression problem as in Section 4.1.2.2. Note that we use the redescending score function in (3.15) to solve the linear regression problem in a robust way. This results (at least asymptotically, for large M) in a Gaussian distributed $\hat{\mathbf{x}}(k|k)$ and the posterior covariance is approximated by (4.29). The likelihood of each filter is evaluated by fitting the innovations to the multivariate density $f(\mathbf{y}(k)|\mathbf{Y}^{k-1}) = \mathcal{N}(\boldsymbol{\nu}_j(k); \mathbf{0}, \mathbf{S}_j(k))$ in (4.30). Given that the linearization errors in (4.9) are small enough this assumption may hold in LOS environments. However, for NLOS environments this is clearly an approximation. Nevertheless, the authors in [5] recommend to use the Gaussian likelihood in non-Gaussian or nonlinear situations even though it does not hold exactly.

The EKF yields precise state estimates in LOS environments whereas the state estimates of the REKF suffer from a higher variance since usefull data is clipped. This leads to rather small innovations $\mathbf{e}_1(k)$ in (4.27) for the EKF and consequently to a high likelihood (4.30). This is because the shape of the multivariate Gaussian pdf to evaluate the likelihood of the EKF is sharper than that of the REKF due to the smaller measurement error covariance $\mathbf{R}_1^*(k)$. In contrast, the spread of the innovations of the REKF is expected to be larger and the innovation covariance (4.28) is higher, resulting in a smaller likelihood (4.30). For NLOS the contrary is true: If outliers occur, a higher spread in the innovations is expected. Since the innovation covariance (4.28) of the EKF is small, the multivariate Gaussian pdf in (4.30) has a sharper shape which leads to small likelihoods $\Lambda_1(k)$. For the REKF the multivariate Gaussian pdf in (4.30) has a flatter shape due to a larger $\mathbf{S}_2(k)$ which leads to higher likelihoods $\Lambda_2(k)$.

At Step 4), the mode probabilities $\mu_j(k)$ for each filter are calculated by using $\Lambda_j(k)$ and normalized such that they sum up to one. In the last step, Step 5), the state and covariance estimates $\hat{\mathbf{x}}_j(k|k)$ and $\mathbf{P}_j(k|k)$ of the two filters are linearly combined with their corresponding mode probability $\mu_j(k)$ yielding the final state and covariance

4.3 Robust Tracking based on M-Estimation and Interacting Multiple Model Algorithm

estimates $\hat{\mathbf{x}}(k|k)$ and $\mathbf{P}(k|k)$, respectively.

Since σ_η^2 is unknown in practice, we suggest replacing it by a multiplicative factor of σ_G^2 to accommodate the higher variance of the NLOS error statistics. If we have uncertainty in the sensor noise variance, e.g., consider that the signal strength decreases when the UE is farther away from the FT leading to a higher variance of the TOA estimates, even though a LOS channel between the FT and the UE exists; then another filter with some presumed measurement covariance can be used in parallel with the two filter developed before. This allows the extended algorithm to accommodate switching of the sensor noise level as well.

Step 1) *Mixing Probability Calculation* $(i, j = 1, 2)$

$$\mu_{i|j}(k-1|k-1) = (1/\bar{c}_j)p'_{ij}\mu_i(k-1) \quad (4.23)$$

$$\bar{c}_j = \sum_i p'_{ij}\mu_i(k-1) \quad (4.24)$$

Step 2) *Interaction* $(j = 1, 2)$

$$\hat{\mathbf{x}}_{0j}(k-1|k-1) = \sum_i \hat{\mathbf{x}}_i(k-1|k-1)\mu_{i|j}(k-1|k-1) \quad (4.25)$$

$$\tilde{\mathbf{x}}_{ij}(k-1|k-1) = \hat{\mathbf{x}}_i(k-1|k-1) - \hat{\mathbf{x}}_{0j}(k-1|k-1)$$

$$\mathbf{P}_{0j}(k-1|k-1) = \sum_i \mu_{i|j}(k-1|k-1)\{\mathbf{P}_i(k-1|k-1)$$

$$+ \tilde{\mathbf{x}}_{ij}(k-1|k-1) \cdot \tilde{\mathbf{x}}_{ij}^\top(k-1|k-1)\} \quad (4.26)$$

Step 3) *Mode-matched Extended Kalman Filtering* $(j = 1, 2)$

$$\hat{\mathbf{x}}_j(k|k-1) = \mathbf{A}\hat{\mathbf{x}}_{0j}(k-1|k-1)$$

$$\mathbf{P}_j(k|k-1) = \mathbf{A}\mathbf{P}_{0j}(k-1|k-1)\mathbf{A}^\top + \mathbf{G}\mathbf{Q}(k)\mathbf{G}^\top$$

$$\mathbf{H}_j(k) = \left.\frac{\partial \mathbf{h}(\mathbf{x}(k))}{\partial \mathbf{x}(k)}\right|_{\mathbf{x}(k) = \hat{\mathbf{x}}_j(k|k-1)}$$

$$\boldsymbol{\nu}_j(k) = \mathbf{y}(k) - \mathbf{h}(\hat{\mathbf{x}}_j(k|k-1)) \quad (4.27)$$

$$\mathbf{S}_j(k) = \mathbf{H}_j(k)\mathbf{P}_j(k|k-1)\mathbf{H}_j^\top(k) + \mathbf{R}_j^*(k) \quad (4.28)$$

For $j = 1$, compute EKF update:

$$\mathbf{K}_1(k) = \mathbf{P}_1(k|k-1)\mathbf{H}_1^\top(k)\mathbf{S}_1^{-1}(k)$$

$$\hat{\mathbf{x}}_1(k|k) = \hat{\mathbf{x}}_1(k|k-1) + \mathbf{K}_1(k)\boldsymbol{\nu}_1(k)$$

$$\mathbf{P}_1(k|k) = (\mathbf{I}_4 - \mathbf{K}_1(k)\mathbf{H}_1(k))\mathbf{P}_1(k|k-1)$$

For $j = 2$, rewrite EKF equations into $\tilde{\mathbf{y}}(k) = \mathbf{D}(k)\mathbf{x}_2(k) + \tilde{\mathbf{v}}(k)$, as in (4.19), where $\mathbf{x}_2(k) = \boldsymbol{\theta}$. Solve for $\boldsymbol{\theta}$ using M-estimation with the algorithm depicted in Figure 3.1 to get $\hat{\mathbf{x}}_2(k|k)$ and

$$\mathbf{P}_2(k|k) = (\mathbf{D}^\top\mathbf{D})^{-1} \quad (4.29)$$

For $j = 1, 2$:

$$\Lambda_j(k) = \mathcal{N}(\boldsymbol{\nu}_j(k); \mathbf{0}, \mathbf{S}_j(k)) \quad (4.30)$$

Step 4) *Mode Probability Update* $(j = 1, 2)$

$$\mu_j(k) = (1/c)\Lambda_j(k)\bar{c}_j, \quad c = \sum_j \Lambda_j(k)\bar{c}_j$$

Step 5) *Combination*

$$\hat{\mathbf{x}}(k|k) = \sum_j \hat{\mathbf{x}}_j(k|k)\mu_j(k), \quad \tilde{\mathbf{x}}_j(k|k) = \hat{\mathbf{x}}_j(k|k) - \hat{\mathbf{x}}(k|k)$$

$$\mathbf{P}(k|k) = \sum_j \mu_j(k) \cdot \left\{\mathbf{P}_j(k|k) + \tilde{\mathbf{x}}_j(k|k) \cdot \tilde{\mathbf{x}}_j^\top(k|k)\right\}$$

Table 4.5: One cycle of the R-IMM algorithm

4.4 Numerical Study

4.4.1 Simulation Environments and Settings

4.4.1.1 Simulation Environments

The positioning accuracy of the trackers is investigated in a network consisting of $M = 5$ FTs illustrated in Figure 4.2. A random trajectory of the UE with starting vector $\mathbf{x}(0) = [1000\text{m} \ 1000\text{m} \ 3\text{m/s} \ 3\text{m/s}]^\mathsf{T}$, is generated according to (4.1) where $\mathbf{Q}(k) = \mathbf{I}_2 \ \forall \ k = 1, 2, \ldots, K$ for each Monte-Carlo run.

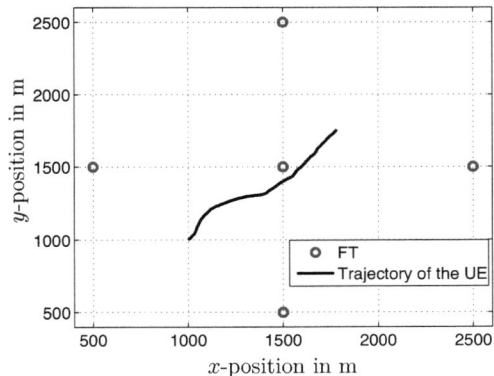

Figure 4.2: Network with $M = 5$ FTs at ($x_1 = 0.5$km, $y_1 = 1.5$km), ($x_2 = 2.5$km, $y_2 = 1.5$km), ($x_3 = 1.5$km, $y_3 = 2.5$km), ($x_4 = 1.5$km, $y_4 = 0.5$km) and ($x_5 = 1.5$km, $y_1 = 1.5$km). The UE is moving on a random trajectory.

Simulations have also been performed for the network treated in Chapter 3 and other network topologies [45]. Here, however, we concentrate on a smaller number of FTs to demonstrate that the noise-adaptive EKF even achieves good performance for small sample sizes.

The sampling time is $\Delta t = 0.2$s, meaning we get $M = 5$ range measurements for positioning each 0.2s. The initial estimate of the UE is $\mathbf{x}(0|0) \sim \mathcal{N}(\mathbf{x}(0), \mathbf{P}(0|0))$ with initial covariance matrix $\mathbf{P}(0|0) = \text{diag}[50^2\text{m}^2 \ 50^2\text{m}^2 \ 6^2(\text{m/s})^2 \ 6^2(\text{m/s})^2]$. Note that the standard deviation of the sensor noise is $\sigma_G = 150$m. Typical values for sensor noise and errors due to NLOS propagation can be found in [92, 111]. For each time step k, the mean error distance (MED) and RMSE, defined in (3.23) and (3.24), are computed to compare the different trackers. In the following, simulations are performed for two different noise models in (4.5) $\mathbf{v}_l(k)$, $l = 1, 2$: The nonlinear model, given in Section 2.2.1.1, where the NLOS occurrences are iid over time and the jump-nonlinear model, given in Section 2.2.1.2, where the NLOS occurrence for each FT changes according to a two-state MC. The parameters of the Markov matrices for different probabilities of NLOS occurrences can be found in Appendix A.2.3. All simulation results are obtained by averaging the performance metrics over 100 Monte-Carlo runs.

4.4.1.2 Settings of the Trackers

The EKF, given in Section 4.1.2.1, and the REKF, explained in Section 4.1.2.2 are computed for comparison purposes. The clipping points for the latter, which uses the redescending score function in (3.15), are set to $c_1 = 1.5$ and $c_2 = 3$. This allows for trading off efficiency in the nominal case (LOS) versus robustness in NLOS environments in an appropriate way (cf. Appendix A.2.1 for a numerical study on the choice of the clipping parameters).

The semi-parametric EKF, presented in Section 4.2, is labeled by EKF-SP-MR because it uses the *modified residuals* approach, given in Figure 3.3, to solve the linear regression problem in the EKF framework. Note that the shape parameter λ for transforming the residuals into a symmetric *approximately* Gaussian sample is obtained by the MLE in (3.22) and the selection is restricted to the interval $0.1 \leq \lambda \leq 1$.

The robust IMM algorithm, presented in Section 4.3, is labeled as R-IMM and uses the EKF with an REKF in parallel. The clipping parameters for the latter are chosen as $c_1 = 0.6$ and $c_2 = 0.8$ to achieve highly robust estimates in severe NLOS environments. Note that an REKF with this parametrization diverges (except if used in parallel with an EKF in an IMM) and loses track in LOS environments because it cuts off too much information. However, used in the IMM algorithm in combination with an EKF provides stability in both LOS and NLOS environments. In this algorithm, the measurement covariance matrix of the EKF is chosen as $\mathbf{R}_1^*(k) = \sigma_G^2 \mathbf{I}_M$ whereas that of the REKF is chosen empirically to $\mathbf{R}_2^*(k) = 3\sigma_G^2 \mathbf{I}_M$. The transition probabilities for the R-IMM are set to $p'_{ij} = 0.5$ $i, j = 1, 2$ because we do not assume any prior knowledge on the switching of the LOS/NLOS events.

For comparison purposes, we choose the competitor from Chen [21, 22] in a KF framework because it does not make any assumptions on the NLOS error statistics. This approach is briefly explained for a stationary UE in Section 3.1.1 and extended in [21] to a moving UE where an KF is used to incorporate a motion model into the positioning algorithm. It is labeled as KF-Rwgh (*KF Residual weighting*) in the sequel. All trackers, except the REKF used in the R-IMM have a measurement covariance of

$\mathbf{R}^*(k) = \sigma_G^2 \mathbf{I}_M$. A summary of the different tracking schemes and their parameters can be found in Table 4.6. The EKF and the KF-Rwgh have lowest computational complexity followed by the robust and semi-parametric trackers which require Cholesky decomposition in (4.18) to compute robust regression estimates, cf. Section 3.3.2.4.

Tracker	Parameters	Reference
EKF	-	Section 4.1.2.1
REKF	$c_1 = 1.5$, $c_2 = 3$	Section 4.1.2.2
SP-MR	λ is set by (3.22), $\lambda \leq 1$, $\hat{\delta} = 1.06\hat{\sigma}_W (\dim(\mathbf{w}))^{-1/5}$	Section 4.2
R-IMM	EKF: $\mathbf{R}_1^*(k) = \sigma_G^2 \mathbf{I}_M$ REKF: $\mathbf{R}_2^*(k) = 3\sigma_G^2 \mathbf{I}_M$, $c_1 = 0.6$, $c_2 = 0.8$	Section 4.3
KF-Rwgh	-	[21, 22]

Table 4.6: Configurations of the different trackers used throughout the simulations.

4.4.2 Simulation Results

4.4.2.1 NLOS Outliers Modeled by a Markov Chain

We first asses the positioning accuracy of the different trackers in an environment where the LOS/NLOS occurrences for each FT are modeled by a two-state Markov chain. Consider that the UE is moving on a 2D plane and a random trajectory, illustrated in Figure 4.2, is generated for each Monte-Carlo run.

Shifted Gaussian Contamination Distribution
The probability of NLOS occurrence equals to $\varepsilon = 25\%$ for each FT and we use a shifted Gaussian distribution with mean $\mu_\eta = 1000$m and standard deviation $\sigma_\eta = 300$m for modeling the NLOS errors. Figure 4.3 illustrates the MED, defined in (3.23), averaged over 100 Monte-Carlo runs versus the discrete time index k for the different trackers. We observe that the EKF completely breaks down since it is not able to cope with large outliers due to NLOS effects. The REKF is able to reduce positioning errors with respect to the EKF and performs similarly to the KF-Rwgh. Both trackers, which underweigh large outliers in different ways, achieve an average MED of approximately 250m. In contrast, EKF-SP-MR adapts to the measurement noise and consequently achieves higher performance, i.e., the time average of the MED is about 80m. Note that the true pdf is difficult to estimate here since it can have two different modes (maxima). It is approximated by the non-parametric pdf estimate

resulting in a small bias in the pdf estimate and consequently to small errors in the state estimates.

The R-IMM achieves highest positioning accuracy because the mode probabilities $\mu_1(k)$ (mode probability for the EKF) and $\mu_2(k)$ (mode probability for the REKF) adapt to the underlying situation. This means that if NLOS FTs are predominant at one time step the spread of the innovations becomes rather large. For the EKF a small likelihood in (4.30) and consequently a small mode probability is calculated; a larger mode probability is then obtained for the REKF. If all FTs are in LOS, the spread of the innovations remains small, yielding a higher mode probability for the EKF and a smaller one for the REKF. Furthermore, since the LOS/NLOS events change according to a Markov chain, the mode probability from the previous time step is used to update the mode probability at the actual time step, improving accuracy of the mode probabilities. Hence, the possibility of adaptation of the mode probabilities to different LOS/NLOS situations and the clipping of large outliers by the redescending score function in the REKF yields superior performance. The final state estimate is either dominated by the REKF with $c_1 = 0.6$ and $c_2 = 0.8$ which consequently discards large outliers and uses the remaining measurements for positioning or by the EKF which takes into account all measurements and consequently yields high precision in LOS environments.

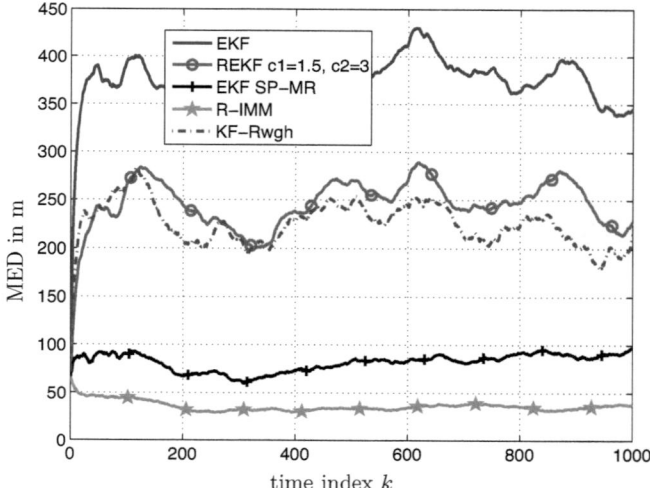

Figure 4.3: MED versus time where $\varepsilon = 0.25$ modeled by a two-state Markov chain for each FT. NLOS errors are Gaussian with $\mu_\eta = 1000$m and $\sigma_\eta = 300$m.

Increasing the percentage of NLOS propagation further up to 50% leads to similar results whereas the positioning accuracy of the R-IMM algorithm rather remains stable compared to the accuracy of the other trackers. This is because the REKF in the R-IMM dominates the state estimates and consequently discards large outliers. However, it is found that the gain in positioning accuracy of the EKF-SP-MR with respect to the KF-Rwgh is reduced to approximately 80m. This comes from the fact that the second mode of the underlying noise pdf becomes more pronounced which complicates non-parametric TKDE for the EKF-SP-MR leading to higher positioning errors. On the other hand, the KF-Rwgh does not estimate any noise pdf but is affected by wrong weighting of the position estimates obtained from the different subgroups. However, for a shifted Gaussian NLOS error pdf and $\varepsilon > 0.6$ the MED of all trackers exceed 500m which makes them inappropriate for such situations.

Exponential Contamination Distribution

Next we consider the NLOS error statistics to be exponentially distributed with standard deviation $\sigma_\eta = 500$m. We investigate the performance of all trackers in a scenario where 50% of the measurements are contaminated by NLOS outliers ($\varepsilon = 0.5$ is the breakdown point of classical M-estimators).

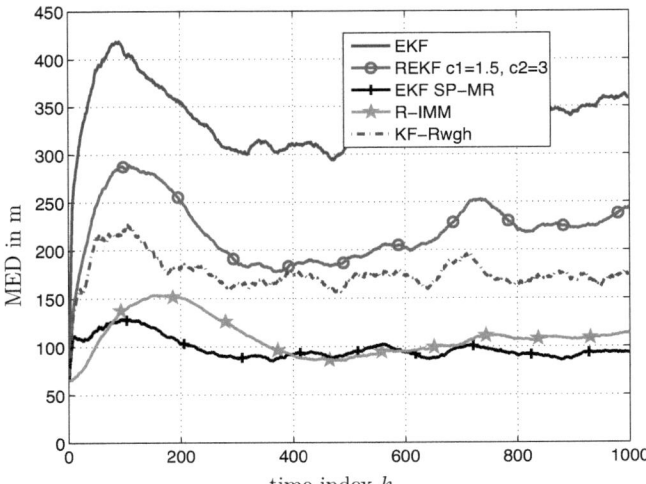

Figure 4.4: MED versus time where $\varepsilon = 0.5$. NLOS occurrences are modeled by a two-state Markov chain for each FT. The contamination pdf is $f_\eta = \mathcal{E}(500\text{m})$.

Figure 4.4 illustrates the MED versus time index k and we can observe again that the EKF breaks down followed by the REKF and KF-Rwgh which gain on average between

120m to 150m positioning accuracy, respectively. The proposed trackers R-IMM and EKF-SP-MR outperform KF-Rwgh by up to 80m. It is interesting to note that the EKF-SP-MR achieves superior performance than its parametric counterpart R-IMM. The reason for this is that it estimates the error distribution non-parametrically yielding a good approximation when the NLOS errors are exponentially distributed. Unlike for a shifted Gaussian pdf f_η, which can result in a bimodal distribution, the errors introduced by an exponential pdf does not usually lead to two modes in the overall distribution. Thus the measurements are well suited for the transformation proposed in (3.21) and pdf estimation gets easier. This results in higher accuracy for the state estimates of EKF-SP-MR. In contrast, the REKF in the IMM algorithm only discards large outliers and consequently loses in performance with respect to EKF-SP-MR. Even though any bounded score function has a breakdown point of at most 50% [51], an REKF or the proposed R-IMM are able to handle situations where slightly more than 50% of the measurements are contaminated by outliers. This is because introducing a motion model in a KF framework decreases the impact of a set of measurements processed at time step k on the final state estimate. Taking into account the previous state estimates together with the actual measurements results in averaging effects that decrease the positioning errors. In contrast, for a stationary UE in Chapter 3, a robust estimator based on the redescending score function (3.15) breaks down when $\varepsilon = 0.5$.

The empirical cdf of the location errors of the different trackers for the same scenario is depicted in Figure 4.5. We observe that for the EKF-SP-MR 95% of the location errors are smaller than 150m whereas for the R-IMM the 95-percentile is at approximately 260m and for the KF-Rwgh it is at 370m. The REKF and the standard EKF further lose in positioning accuracy.

Increasing ε further makes the limitations of the R-IMM tracker apparent. Figure 4.6 illustrates the error cdf of a scenario where 75% of the range measurements are contaminated by NLOS errors modeled by an exponential pdf with $\sigma_\eta = 500$m. While the EKF-SP-MR performs best and yields a 95-percentile of approximately 200m, KF-Rwgh loses about 100m and the R-IMM loses 300m for the same error percentile. The reason for this is that the former ones exploit all measurements whereas the R-IMM sometimes discards information by clipping of the observations. Even though the R-IMM yields smaller errors on average than the EKF, its 95-percentile is similar to the one of the REKF at approximately 580m. It is interesting to note that large positioning errors for the R-IMM occur with low frequency leading to a slow convergence of its error cdf to one which corresponds to divergence of the tracker in these cases. Since the innovation sequences are rather large, the REKF within the R-IMM filter is dominant, meaning $\mu_2(k)$ tends to one. Then, the small clipping points for the REKF within the R-IMM result in cutting too much information and in some cases to divergence of the filter.

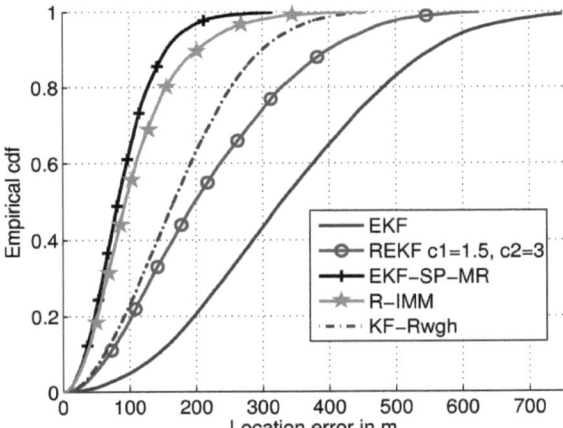

Figure 4.5: Empirical cdf of location errors where $\varepsilon = 0.5$ modeled by a two-state Markov chain for each FT. The contamination pdf f_η is modeled by an exponential distribution with $\sigma_\eta = 500$m.

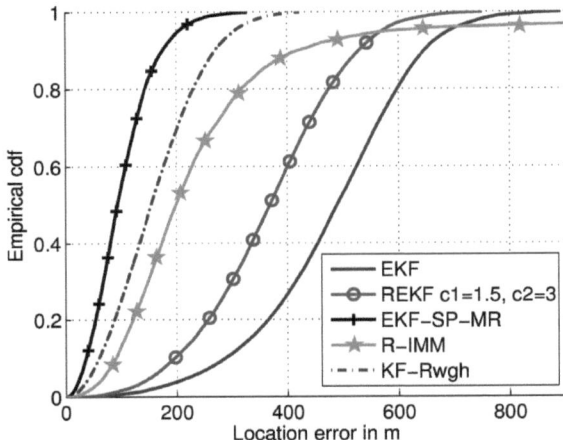

Figure 4.6: Empirical cdf of location errors where $\varepsilon = 0.75$ modeled by a two-state Markov chain for each FT. The contamination pdf f_η is modeled by an exponential distribution with $\sigma_\eta = 500$m.

In general the noise realizations from the exponential pdf with $\sigma_\eta = 500$m are usually smaller than those of a shifted Gaussian error pdf with $\mu_\eta = 1000$m and $\sigma_\eta = 300$m, leading in general to higher positioning accuracy when the NLOS errors are exponentially distributed.

In order to investigate the behaviour of the trackers to changing noise variances we study the MED in terms of the standard deviation of the NLOS error statistics σ_η for a scenario where 50% of the measurements are contaminated by NLOS outliers, modeled as an exponential distribution. Simulation results are illustrated in Figure 4.7. First, we observe that the higher the standard deviation of the NLOS errors the higher the positioning errors of all trackers become. The EKF completely breaks down when σ_η increases whereas the REKF confers some robustness on the state estimates but still yields large positioning errors. The reason for this is that the clipping parameters c_1 and c_2 of the latter are chosen to trade off efficiency versus robustness.

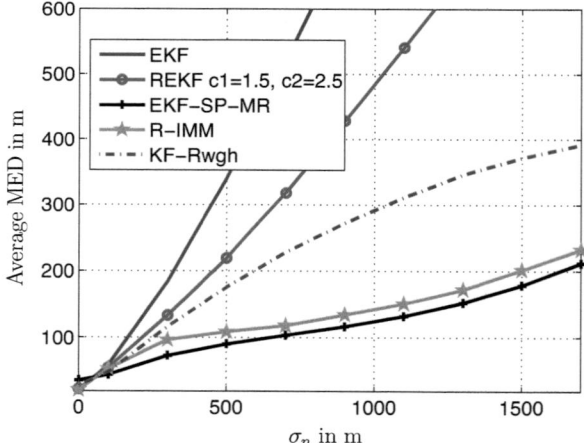

Figure 4.7: Average MED versus the standard deviation of NLOS error statistics where $\varepsilon = 0.5$ modeled by a two-state Markov chain for each FT. An exponential contamination pdf is used for modeling NLOS errors.

A different choice of these parameters (with smaller c_1 and c_2) is carried out in the REKF used in the R-IMM algorithm which consequently achieves higher accuracy since large measurements are discarded. However, for large σ_η, the tracker KF-Rwgh behaves more stable than REKF because, for the former, large outliers have less impact on the state estimates. The proposed algorithms, R-IMM and EKF-SP-MR, significantly outperform their competitors and gain up to 200m in positioning accuracy with respect to KF-Rwgh. Note that the EKF-SP-MR gains approximately 20m with respect to R-IMM for various values of σ_η. While the R-IMM algorithm explicitly incorporates the switching into the mode probabilities $\mu_{1,2}(k)$, the EKF-SP-MR does not make any assumptions on the switching of noise sequences and estimates the noise pdf at each time step k to exploit all moments of the noise statistics. Thus, for non-Gaussian f_η incorporation of higher order moments seems to play a more dominant role for the final state estimate than incorporating transition probabilities.

In contrast, the R-IMM gains in performance with respect to EKF-SP-MR if f_η is a shifted Gaussian pdf for $\varepsilon < 0.6$. This is because it is implicitly based on the Gaussian assumption and cuts off large outliers whereas the remaining zero-mean Gaussian sensor noise is in favor of state estimation.

4.4.2.2 Observations are iid

Gaussian Sensor Noise Only

The question now arises how the different trackers perform in the nominal case, meaning when all FTs are in LOS. An REKF with clipping parameters $c_1 = 0.6$ and $c_2 = 0.8$ is added to demonstrate the performance loss in the nominal case. Figure 4.8 depicts the MED versus time where all FTs are in LOS, i.e., $\varepsilon = 0$.

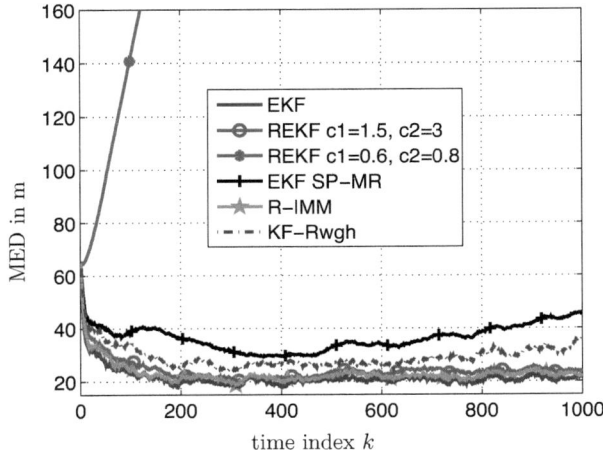

Figure 4.8: MED versus time index k in LOS environments ($\varepsilon = 0$).

We observe that the EKF achieves highest accuracy followed by the R-IMM. This is due to overweighing the EKF in the R-IMM because the spread of the innovations is rather small and they well fit the innovation covariance $\mathbf{S}_1(k)$ resulting in a high likelihood $\Lambda_1(k)$ for different k. On the other hand, since $\mathbf{R}_2^*(k) = 3\mathbf{R}_1^*(k)$ the likelihood of the REKF $\Lambda_2(k)$ inside the R-IMM gets small. Consequently the mode probability $\mu_1(k)$ tends to one for all time steps and the performance of the R-IMM is similar to the performance of the EKF. The tracker REKF with $c_1 = 1.5$ and $c_2 = 3$ only slightly loses with respect to the EKF followed by KF-Rwgh and EKF-SP-MR which lose between 10m and 20m, respectively. Note that the good performance of the R-IMM is based on the fact that the state estimates are dominated by the EKF. Since the clipping points of the REKF are $c_1 = 0.6$ and $c_2 = 0.8$ highly robust estimates in NLOS environments can be achieved at the expense of larger errors when ε decreases (confer Appendix A.2.1). The MED of an REKF with these parameters increases to more than 400m in LOS environments and the tracker diverges which can be observed in Figure 4.8. Instead, the R-IMM algorithm intelligently combines the states estimates of both trackers yielding to high efficiency in LOS environments and to robustness in NLOS environments. Note that increasing the number of FTs to, e.g. $M = 10$, decreases the difference in performance between the EKF-SP-MR and the EKF to about 8m because non-parametric estimation of the noise pdf is simplified when the number of observations increases.

Influence of Probability of NLOS occurrence
Last, we study the impact of the probability of NLOS occurrence on the positioning accuracy where $f_\eta = \mathcal{E}(500\text{m})$. Figure 4.9 illustrates the MED averaged over time and Monte-Carlo runs versus ε and we observe that as soon as ε differs from zero the EKF significantly loses in positioning accuracy.
The parametric trackers REKF and R-IMM perform best for small ε whereas the former loses when ε exceeds 0.2, and the latter behaves robustly up to $\varepsilon = 0.5$. Increasing ε further leads to large errors. At some ε even the information on the previous state estimate in the R-IMM and the measurements do not contain enough information to hold the track. Note that, if we use the R-IMM in iid environments it loses performance with respect to the switching case since the algorithm incorporates information on the mode probabilities from the previous time step in the present one. This results in a mismatch and higher positioning errors are expected.
In contrast, the semi-parametric tracker loses in positioning accuracy with respect to its competitors for small ε. However, when ε exceeds 30% it significantly outperforms all trackers and gains up to 50m in positioning accuracy with respect to KF-Rwgh.

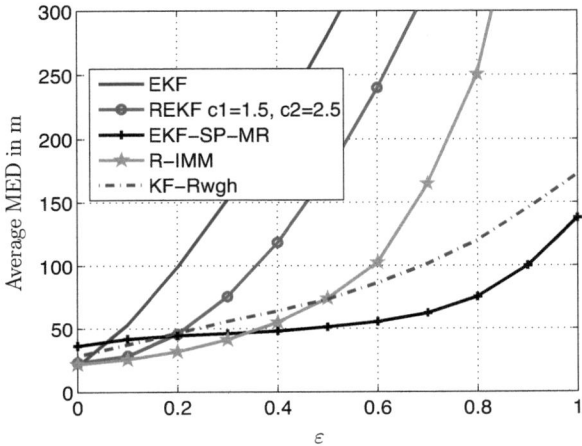

Figure 4.9: Average MED versus ε. The contamination pdf is $f_\eta = \mathcal{E}(500\text{m})$ and NLOS errors are iid.

4.5 Discussion and Conclusions

In this Chapter, a parametric and a semi-parametric tracking algorithm to mitigate errors due to NLOS propagation have been proposed. Both approaches use either robust or semi-parametric techniques from time series analysis within a Bayesian framework. The former (R-IMM) is based on theory of robust statistics. In this context, designing an estimator, always involves the following question: *How much efficiency loss are we willing to pay for the nominal case (LOS) to obtain a certain degree of robustness in NLOS environments?* Efficiency in LOS and robustness in NLOS are contradictory objectives since it is not possible to achieve both at the same time. A solution to this problem is proposed by designing a robust IMM algorithm that uses two filters in parallel. One of the filters, the EKF, is well suited for LOS situations whereas the second filter, a robustified EKF, is adapted to situations where a large amount of NLOS errors are expected. The state estimates from the filter with the highest likelihood is overweighed. It is shown via numerical simulations that the proposed algorithm yields positioning accuracy similar to the EKF in LOS environments and significantly outperforms different competitors for small to medium percentages of NLOS errors.

Since the IMM structure models random switching of LOS/NLOS events it is preferable to be used in such situations rather than in iid environments where no switching occurs.

Even though robust M-estimators in the framework of *robust statistics* are robust to at most $\varepsilon = 50\%$ of contamination the proposed R-IMM algorithm achieves acceptable positioning accuracy in situations where this percentage is slightly exceeded. This gain in performance is thanks to the motion model which is incorporated within a Bayesian framework and consequently yields higher stability. However, the limitations of the R-IMM tracker become apparent when the percentage of NLOS outliers increases beyond 60% and the magnitude of the outliers are large with respect to sensor noise. In this case large positioning errors and a loss in performance with respect to KF-Rwgh can be observed.

In contrast, the semi-parametric tracker (EKF-SP-MR) uses the parametric form of the EKF equations, that are reformulated into a linear regression problem, and the interference pdf of this model at each time step is estimated non-parametrically. This pdf estimate is incorporated into the update step to yield the position estimate. Simulation results show that this tracker achieves robustness for non-Gaussian interference pdfs irrespective of whether the NLOS occurrences are modeled as iid or according to a Markov chain. Even for a small number of FTs good performance in both LOS and NLOS environments is obtained. Increasing the number of FTs results in further improvements since non-parametric pdf estimation is simplified. The main limitation of this tracker is the loss of about 14m positioning accuracy in the nominal LOS case when only a small number of range measurements are available.

In situations where NLOS FTs are predominant the EKF-SP-MR achieves better performance than R-IMM and various competing methods for non-Gaussian f_η because it incorporates higher order moments of the interference pdf for state estimation and consequently improves performance. In contrast, the R-IMM gains in positioning accuracy when f_η is a shifted Gaussian pdf. In this case, large outliers are discarded and the R-IMM, which is implicitly based on the Gaussian assumption, uses the remaining measurements, containing only Gaussian sensor noise, for state estimation.

For the time being, the posterior covariance matrix of the EKF-SP-MR is approximated by the posterior covariance of a standard EKF. Performance in terms of the positioning accuracy can be improved if this matrix is estimated more accurately. For this purpose, the covariance of the semi-parametric estimator has to be derived.

Since the measurement covariance matrix is unknown in general, adaptive estimation of this quantity can result in higher positioning accuracy for the R-IMM algorithm. While for a white noise sequence a running variance estimator may be suitable, it fails for the jump-nonlinear model because the time steps where the measurement noise covariance changes are unknown. Hence, the latter requires additional tracking of the LOS/NLOS events or more sophisticated covariance estimation schemes.

Chapter 5
Tracking based on Outlier Detection and Data Association

Here, the same problem as in Chapter 4 is treated, namely tracking a UE based on TOA range measurements in mixed LOS/NLOS environments. Instead of using robust and semi-parametric techniques that take into account all measurements for positioning, we propose a joint outlier detection and tracking scheme where large outliers are discarded. For this purpose N different subgroups of range measurements are constructed to compute N position estimates via LS with their corresponding covariance matrices. Both quantities of each subgroup are used in a parametric hypothesis test for NLOS detection. The accepted position estimates are weighted with different probabilities in a Kalman filter framework whereas the rejected ones are discarded. Simulation results show a significant increase in positioning accuracy in NLOS environments with respect to different robust competing trackers. In LOS environments similar performance to the EKF is achieved. The method developed hereafter does not assume any statistical knowledge of the NLOS errors and only assumes the sensor noise variance to be known. The proposed approach, published in [47], is termed modified probabilistic data association (MPDA) and relates to the probabilistic data association (PDA) algorithm [62] developed for target tracking in radar in the presence of measurement uncertainties.

5.1 Problem Statement

5.1.1 Signal Model

Recall the signal model from Section 4.1.1 where a UE with state vector $\mathbf{x}(k) = [x(k)\ y(k)\ \dot{x}(k)\ \dot{y}(k)]^\mathsf{T}$ is moving on a random trajectory and TOA range measurements are observed in $\mathbf{y}(k)$. For the sake of readability the state space equations for time step k are rewritten

$$\mathbf{x}(k) = \mathbf{A}\mathbf{x}(k-1) + \mathbf{G}\boldsymbol{\omega}(k-1), \tag{5.1}$$
$$\mathbf{y}(k) = \mathbf{h}(\mathbf{x}(k)) + \mathbf{v}_l(k), \qquad l = 1, 2, \tag{5.2}$$

where the quantities are defined in Section 4.1.1. Recall that two different measurement noise models are available for the noise process $\mathbf{v}_l(k)$. It either is modeled as iid ($l = 1$) or it is modeled as a two-state Markov chain ($l = 2$) for each FT.

5.1.2 Context of Research and Existing Methods

Suboptimal robust and non-robust approaches to the tracking problem such as the EKF and the REKF have been presented in Section 4.1.2 and serve for comparison purposes here. In some cases, further improvements can be achieved using semi-parametric methods [45], given in Section 4.2, where the interference pdf at each time step k is estimated in an EKF framework taking into account all measurements. Then, this pdf estimate is used to determine the position of the UE based on the ML principle.

While these techniques use all observations for positioning (even though the REKF may clip off measurements by using the redescending score function in Section 3.2.2), the method presented hereafter only takes into account observations that are likely to come from LOS FTs. In this context the authors in [88] consider a Kalman filter in combination with an outlier detection scheme based on a likelihood ratio test. This filter discards erroneous measurements and only uses the remaining ones for the update step. When no measurements are accepted in the test, the predicted state and covariance estimates are used in the update step. A similar approach is suggested in [4] for the NLOS problem where a moving UE is considered and the detected NLOS measurements are excluded in the update step of the UKF. Other detection techniques are applied to NLOS identification in [83, 91, 101, 111] where each FT is tested individually. Usually the detected NLOS measurements are discarded for calculating the position estimate, meaning a hard-decision approach is performed and the remaining range measurements are processed to yield the position estimate.

Here, we propose a different method relating to approaches such as [18, 21, 22, 40] that construct different subgroups of TOA range measurements which are combined to yield the final position estimate. Chen suggests a residual weighting algorithm [21, 22] which requires computation of the weighted residuals from LS estimates calculated from the different subgroups. The final location estimate is the linear combination of the LS position estimates weighted inversely to their residuals. This approach takes into account all measurements, meaning it softly combines the confidence one has on a particular subgroup of FTs. In contrast, the authors in [18, 40] suggest hard-decision approaches, meaning the NLOS FTs are identified and discarded for positioning. In [18] the *least median of squares* is computed to yield the final position estimate whereas in [40] a different criterion is chosen to select the most reliable FTs. The approaches are explained in more detail in Section 3.1.1. All references mentioned above avoid imposing a specific NLOS error distribution since it is unknown in practice.

Here, our contribution is a combination of a hard- and soft-decision approach which uses the advantages of both, meaning large outliers are discarded and the accepted data is weighted with different probabilities. First, N different subgroups are constructed and N different position estimates with their corresponding covariance matrix are computed by a LS algorithm. The position estimate with the smallest noise realization and the best geometric constellation is preferable for doing the update. However, we neither know the particular noise realization (otherwise we could subtract it) nor do we know the best geometric constellation since different constellations appear similarly well suited for estimating the position of the UE. Therefore we need to detect the best position estimate from a subgroup of FTs. A hard-decision approach, that discards all but one position estimate, is not suitable because useful information is lost in scenarios where LOS FTs are predominant. On the other hand, weighting all measurements without discarding large outliers can still lead to inaccurate estimates when the weights are not chosen appropriately.

Therefore, unlike in [18, 21, 22, 40], we propose to test all position estimates, reject the ones computed by NLOS FTs, and assign different probabilities for the remaining position estimates similar to the idea of PDA [62]. Our approach does not assume any statistical knowledge of the NLOS errors and only assumes the sensor noise variance to be known. Without loss of generality, the proposed positioning algorithm uses TOA measurements but can be easily extended to other signal parameters such as TDOA, RSS or AOA. For TOA measurements, the proposed algorithm is only applicable in situations where more than three FTs are available.

5.2 Modified Probabilistic Data Association

5.2.1 General Concept

The flowchart of the MPDA algorithm is illustrated in Figure 5.1. First, we construct N different subgroups of range measurements $\mathbf{y}(k)$ together with the positions of the corresponding FTs. Each subgroup is used to compute the LS position estimates $\mathbf{z}_n(k) = [\hat{x}_n(k)\ \hat{y}_n(k)]^\mathsf{T}$, $n = 1, 2, \ldots, N$ (*least-squares estimation*). The corresponding position estimate has high accuracy if the FTs in subgroup n are in LOS since the LS estimator is optimal for Gaussian noise. However, as soon as one FT is contaminated by an NLOS outlier, LS estimation breaks down and the RMSE of $\mathbf{z}_n(k)$ increases. This effect can be used to test each position estimate of the N subgroups against LOS. The advantage of the grouping is that it allows to incorporate additional information, meaning the geometric constellation of the FTs into the hypothesis test. This is done by using the covariance matrix of the position estimates $\mathsf{Cov}\{\mathbf{z}_n(k)\}$ which contains the positions of the FTs from subgroup n. In contrast, the geometric information is not used when range measurements are tested for NLOS occurrence, based on methods like the running mean or running variance [10, 91, 111].

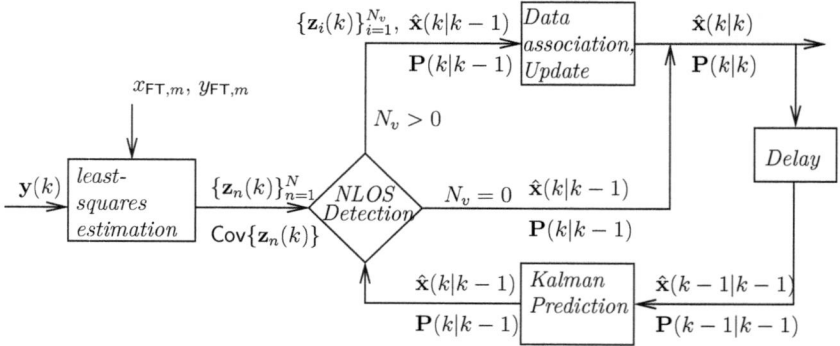

Figure 5.1: Flow chart for one cycle of the MPDA algorithm.

The hypothesis test in the proposed MPDA algorithm, denoted as *NLOS Detection*, also requires the predicted state and covariance matrix computed from the previous time step $k-1$ in *Kalman Prediction*. If the number of accepted position estimates N_v is greater than zero ($N_v > 0$) the position estimates are associated with different probabilities (*Data association, Update*) to yield the final state estimate $\hat{\mathbf{x}}(k|k)$ and its covariance $\mathbf{P}(k|k)$. In the ideal case, meaning when all M FTs are in LOS, all $\mathbf{z}_n(k)$ are accepted with a certain probability $\mathsf{P_G}$ that has to be set a priori. Then, the subgroup with the best geometric constellation (or smallest error covariance) is overweighed by the Kalman filter. If none of the LS position estimates are accepted in the hypothesis test ($N_v = 0$), the predicted state and covariance are used instead.

5.2.2 Grouping and Positioning

Since at least $M = 3$ range measurements are needed to determine the position of a UE, we have $N = \binom{M}{3}$ different subgroups of range measurements which are used together with the positions of the corresponding FTs to determine the position of the UE. Here, the position estimates are computed using a Gauss-Newton algorithm [11]. Assume that the Gauss-Newton algorithm is unbiased in LOS environments (this is not true in general, but numerical simulation results have shown that the bias is negligible). Then, we have the new measurement equation

$$\mathbf{z}_n(k) = \mathbf{B}\mathbf{x}(k) + \boldsymbol{\xi}_n(k) \qquad n = 1, \ldots, N, \quad \text{with} \quad \mathbf{B} = \begin{bmatrix} 1 & 0 & 0 & 0 \\ 0 & 1 & 0 & 0 \end{bmatrix} \qquad (5.3)$$

where

$$\boldsymbol{\xi}_n(k) \sim \mathcal{N}(\mathbf{0}, \sigma_G^2 (\mathbf{H}_n^\mathsf{T}(k)\mathbf{H}_n(k))^{-1}) \qquad n = 1, \ldots, N \qquad (5.4)$$

5.2 Modified Probabilistic Data Association

if all FTs from subgroup n are in LOS, where $\mathbf{H}_n(k)$ is the Jacobian matrix from subgroup n evaluated at $\mathbf{z}_n(k)$ containing the positions of the corresponding FTs. To prevent numerical problems it should be avoided to choose a set of FTs that are located on a straight line. In this case the geometric constellation is disadvantageous which leads to ambiguities in finding the position of the UE. Then, the matrix $(\mathbf{H}_n^\mathsf{T}(k)\mathbf{H}_n(k))$ can be singular and is thus not invertible.

The starting point of the Gauss-Newton algorithm can be calculated for instance by a computational inexpensive one-step algorithm [106]. However, (5.3) does not hold anymore if one or more FTs from subgroup n are in NLOS. Thus, a test is required to detect whether $\mathbf{z}_n(k)$ is computed from a subgroup that contains at least one NLOS measurement to discard the corresponding position estimate $\mathbf{z}_n(k)$. Then, the remaining position estimates can be used in a Kalman filter framework where different probabilities are assigned to the different measurements.

5.2.3 Algorithm

The different steps from the MPDA algorithm illustrated in Figure 5.1 are explained in more detail in the sequel.

5.2.3.1 Kalman Prediction

For initializing the Kalman filter we assume that $\hat{\mathbf{x}}(0|0) \sim \mathcal{N}(\mathbf{x}(0), \mathbf{P}(0|0))$ which can be computed e.g. by LS. The time update is performed for $k = 1, \ldots, K$ as in a conventional Kalman filter, i.e,

$$\hat{\mathbf{x}}(k|k-1) = \mathbf{A}\hat{\mathbf{x}}(k-1|k-1) \tag{5.5}$$

$$\mathbf{P}(k|k-1) = \mathbf{A}\mathbf{P}(k-1|k-1)\mathbf{A}^\mathsf{T} + \mathbf{G}\mathbf{Q}\mathbf{G}^\mathsf{T} \tag{5.6}$$

$$\hat{\mathbf{z}}(k|k-1) = \mathbf{B}\hat{\mathbf{x}}(k|k-1) \tag{5.7}$$

$$\boldsymbol{\nu}_n(k) = \mathbf{z}_n(k) - \hat{\mathbf{z}}(k|k-1) \quad n = 1, \ldots, N, \tag{5.8}$$

5.2.3.2 NLOS Detection

From (5.3) and (5.8) it can be deduced that

$$\boldsymbol{\nu}_n(k) \sim \mathcal{N}(\mathbf{0}, \mathbf{S}_n(k)) \quad n = 1, 2, \ldots, N, \tag{5.9}$$

holds if all FTs at time step k are in LOS where the innovation covariance matrix for the position estimate from subgroup n is

$$\mathbf{S}_n(k) = \mathbf{BP}(k|k-1)\mathbf{B}^\mathsf{T} + \sigma_G^2(\mathbf{H}_n^\mathsf{T}(k)\mathbf{H}_n(k))^{-1} \quad n = 1, \ldots, N. \quad (5.10)$$

In order to validate (5.9), we define the following N hypotheses and alternatives:

$$\mathsf{H}_{0,n}: \quad \boldsymbol{\nu}_n(k) \sim \mathcal{N}(\mathbf{0}, \mathbf{S}_n(k)) \quad n = 1, 2, \ldots, N \quad (5.11)$$

$$\mathsf{H}_{1,n}: \quad \text{not } \mathsf{H}_{0,n} \quad n = 1, 2, \ldots, N. \quad (5.12)$$

The hypothesis $\mathsf{H}_{0,n}$ holds true, if the FTs from subgroup n are in LOS. The alternative $\mathsf{H}_{1,n}$ holds if at least one FT from subgroup n is contaminated by NLOS errors. Then, $\mathbf{z}_n(k)$ is erroneous and a higher sample covariance of the innovations (5.8) is expected, yielding the mismatch. Note that the Gaussian assumption of the innovations (5.9) is not fulfilled in NLOS environments.

If $\mathbf{z}_n(k)$ is computed from LOS FTs, it falls in a certain region (validation gate) around the predicted measurement. To evaluate this, we compute the test statistic $T_n(k)$ as

$$T_n(k) = \boldsymbol{\nu}_n^\mathsf{T}(k)\mathbf{S}_n^{-1}(k)\boldsymbol{\nu}_n(k), \quad n = 1, 2, \ldots, N, \quad (5.13)$$

and compare it with the threshold γ, based on a preset false alarm rate $\mathsf{P}_{\mathsf{FA}} = 1 - \mathsf{P}_{\mathsf{G}}$ under assumption (5.9). The distribution of $T_n(k)$ under $\mathsf{H}_{0,n}$ is approximated by the chi-squared distribution with two degrees of freedom [5]. If $T_n(k)$ is larger than the threshold γ the hypothesis $\mathsf{H}_{0,n}$ is rejected, otherwise it is retained. Then, the probability P_{G} that a measurement from LOS FTs falls in the validation gate is

$$\int_0^\gamma f_{\chi^2(2)}(x)dx = \mathsf{P}_{\mathsf{G}} = 1 - \mathsf{P}_{\mathsf{FA}}, \quad (5.14)$$

where $f_{\chi^2(2)}(\cdot)$ is the chi-squared pdf with two degrees of freedom. The test is performed for each $\mathbf{z}_n(k)$ and the accepted ones are weighted with different probabilities, explained in Section 5.2.3.3, before doing the Kalman update, given in Section 5.2.3.4. Note that the validated measurements are labeled as $\mathbf{z}_j(k)$, $j = 1, 2, \ldots, N_v(k)$ where $N_v(k)$ is the total number of validated measurements.

However, if none of the measurements $\mathbf{z}_n(k)$ is accepted in the tests, the predicted state and covariance estimates of the standard Kalman filter from (5.5) and (5.6), respectively, are used as the final state estimates.

5.2.3.3 Data Association

At each time step k only one position estimate $\mathbf{z}_j(k)$ is assumed to be the correct one. This is the one estimated from a subgroup of LOS FTs resulting in the smallest

5.2 Modified Probabilistic Data Association

error covariance $\sigma_G^2(\mathbf{H}_j^{\mathsf{T}}(k)\mathbf{H}_j(k))^{-1}$ which inherently incorporates the geometric constellation of the FTs. The other validated measurements are assumed to be "clutter" (uniformly distributed in the validation gate) calculated from FTs with less advantageous geometric constellations or higher sensor noise realizations. Instead of deciding which measurements best fits the model, we propose a soft decision metric where the accepted measurements are weighted with different probabilities. In order to calculate the probability that each validated measurement is the correct one, we define the following events, following the approach taken in [62]:

E_j : $\{\mathbf{z}_j(k)$ is the correct position estimate; determined from LOS FTs with smallest error covariance, $j = 1, \ldots, N_v(k)$ $\}$

E_0 : $\{$none of the position estimates at time k stems from LOS FTs$\}$

The association probabilities are

$$\beta_j(k) = \mathsf{Pr}\{E_j(k)|\mathbf{Z}^k\} \qquad (5.15)$$
$$= \mathsf{Pr}\{E_j(k)|\mathbf{Z}(k), N_v(k), \mathbf{Z}^{k-1}\}, \qquad j = 0, 1, \ldots, N_v(k), \qquad (5.16)$$

where $\mathbf{Z}(k) = \{\mathbf{z}_j(k)\}_{j=1}^{N_v(k)}$ and $\mathbf{Z}^k$ is the cumulative set of measurements, i.e., $\mathbf{Z}^k = \{\mathbf{Z}(i)\}_{i=1}^{k}$.

Now we assume that the innovations $\boldsymbol{\nu}_j(k)$, $j = 1, \ldots, N_v(k)$ are mutually independent. This is true when we construct N mutually exclusive subgroups in Section 5.2.2 instead of considering $\binom{M}{3}$ subgroups. For the derivation of the association probabilities we assume that the assumption of mutually independent subgroups holds. Using Bayes rule, (5.15) can be rewritten as

$$\beta_j(k) = \frac{1}{c} f(\mathbf{Z}(k)|E_j(k), N_v(k), \mathbf{Z}^{k-1}) \mathsf{Pr}\{E_j(k)|N_v(k), \mathbf{Z}^{k-1}\} \quad j = 0, 1, \ldots, N_v(k), \qquad (5.17)$$

where c is a normalization constant and $f(\cdot)$ is the joint density of the accepted/rejected measurements conditioned on $E_j(k), N_v(k), \mathbf{Z}^{k-1}$. For E_j, $j \neq 0$ it is the product of the assumed Gaussian pdf of the correct position estimate and the pdfs of the incorrect (accepted) position estimates uniformly distributed in the validation region. The pdf of the correct measurement is a truncated normal distribution,

$$f(\mathbf{z}_j(k)|E_j(k), N_v(k), \mathbf{Z}^{k-1}) = \mathsf{P}_G^{-1} \mathcal{N}(\mathbf{z}_j(k); \hat{\mathbf{z}}(k|k-1), \mathbf{S}_j(k)) \qquad (5.18)$$
$$= \mathsf{P}_G^{-1} \mathcal{N}(\boldsymbol{\nu}_j(k); \mathbf{0}, \mathbf{S}_j(k)) \qquad (5.19)$$
$$= \mathsf{P}_G^{-1} \frac{\exp\{-\frac{1}{2}\boldsymbol{\nu}_j^{\mathsf{T}}(k)\mathbf{S}^{-1}(k)\boldsymbol{\nu}_j(k)\}}{2\pi|\mathbf{S}_j(k)|^{0.5}}, \qquad (5.20)$$

where $|\mathbf{S}_j(k)|$ denotes the determinant of matrix $\mathbf{S}_j(k)$. The first factor on the right hand side from (5.17) is

$$f(\mathbf{Z}(k)|E_j(k), N_v(k), \mathbf{Z}^{k-1}) = \begin{cases} \prod_{i=1, i\neq j}^{N_v(k)} \frac{\mathcal{N}(\boldsymbol{\nu}_j(k); \mathbf{0}, \mathbf{S}_j(k))}{\mathcal{V}_i(k) \mathsf{P}_\mathsf{G}}, & j = 1, \ldots, N_v(k) \\ \prod_{i=1}^{N_v(k)} \mathcal{V}_i^{-1}(k), & j = 0, \end{cases} \quad (5.21)$$

where $\mathcal{V}_j(k)$ is the area of the validation region [62] of the $N_v(k)$ accepted hypothesis, i.e.,

$$\mathcal{V}_j(k) = \gamma \pi |\mathbf{S}_j(k)|^{0.5} \quad j = 1, 2, \ldots, N_v(k). \quad (5.22)$$

Hence, in (5.21) it is assumed that the correct measurement is Gaussian distributed around $\hat{\mathbf{z}}(k|k-1)$ and the other measurements are uniformly distributed in their corresponding validation region $\mathcal{V}_j(k)$. Note that a precise measurement from an advantageous geometric constellation of FTs results in a small innovation covariance $\mathbf{S}_j(k)$ which yields a high likelihood $\mathcal{N}(\boldsymbol{\nu}_j(k); \mathbf{0}, \mathbf{S}_j(k))$. Consequently a high probability $\beta_j(k)$ is assigned to the corresponding innovation $\boldsymbol{\nu}_j(k)$ which is then overweighed with respect to the others in the update step, in Section 5.2.3.4, explained hereafter. In contrast, if the covariance $\mathbf{S}_j(k)$ is rather high, the likelihood decreases yielding a smaller β_j.

The prior probabilities from (5.17) are

$$\Pr\{E_j(k)|N_v(k), \mathbf{Z}^{k-1}\} = \begin{cases} \frac{\mathsf{P}_\mathsf{G}}{N_v(k)} & j = 1, \ldots, N_v(k) \\ 1 - \mathsf{P}_\mathsf{G} & j = 0. \end{cases} \quad (5.23)$$

Multiplying the prior with the pdf of the correct measurements yields

$$\beta'_j = \prod_{i=1, i\neq j}^{N_v(k)} \mathcal{V}_i^{-1}(k) \mathsf{P}_\mathsf{G}^{-1} \frac{\mathsf{P}_\mathsf{G}}{N_v(k)} \mathcal{N}(\boldsymbol{\nu}_j(k); \mathbf{0}, \mathbf{S}_j(k)), \quad j = 1, \ldots, N_v(k) \quad (5.24)$$

$$\beta'_0 = \prod_{i=1}^{N_v(k)} \mathcal{V}_i^{-1}(k)(1 - \mathsf{P}_\mathsf{G}), \quad j = 0. \quad (5.25)$$

To ensure that $\sum_{j=0}^{N_v(k)} \beta_j = 1$ we normalize β'_j as

$$\beta_j = \frac{\beta'_j}{\beta'_0 + \sum_{i=1}^{N_v(k)} \beta'_i} \quad \forall\, j = 0, 1, \ldots, N_v(k). \quad (5.26)$$

Note that if no measurement is accepted in the test, β_0 yields one.

5.2.3.4 Update

In Section 5.2.3.3, we define different innovation covariance matrices $\mathbf{S}_n(k)$ for the position estimates computed from different subgroups of FTs. In order to calculate the

Kalman gain, we approximate the innovation covariance matrix as

$$\mathbf{S}(k) = \mathbf{B}\mathbf{P}(k|k-1)\mathbf{B}^\mathsf{T} + \sigma_G^2 \mathbf{I}_2, \tag{5.27}$$

where $\mathbf{I}_2$ is a 2×2 identity matrix meaning we assume a measurement covariance matrix that has the same resolution in x- and y-direction. The Kalman gain is then

$$\mathbf{K}(k) = \mathbf{P}(k|k-1)\mathbf{B}^\mathsf{T}\mathbf{S}^{-1}(k). \tag{5.28}$$

The final state estimate is determined using the weighted innovations [62], i.e.,

$$\hat{\mathbf{x}}(k|k) = \hat{\mathbf{x}}(k|k-1) + \mathbf{K}(k) \sum_{j=1}^{N_v(k)} \beta_j(k)\boldsymbol{\nu}_j(k). \tag{5.29}$$

The covariance update is calculated as in [62], thus

$$\mathbf{P}(k|k) = \beta_0(k)\mathbf{P}(k|k-1) + (1-\beta_0(k))\mathbf{P}_c(k|k) + \tilde{\mathbf{P}}(k), \tag{5.30}$$

where

$$\mathbf{P}_c(k|k) = (\mathbf{I}_4 - \mathbf{K}(k)\mathbf{B}(k))\mathbf{P}(k|k-1), \tag{5.31}$$

is the posterior covariance matrix of the standard Kalman filter and

$$\tilde{\mathbf{P}}(k) = \mathbf{K}(k) \left[\sum_{j=1}^{N_v(k)} \beta_j(k)\boldsymbol{\nu}_j(k)\boldsymbol{\nu}_j^\mathsf{T}(k) - \boldsymbol{\nu}(k)\boldsymbol{\nu}^\mathsf{T}(k) \right] \mathbf{K}^\mathsf{T}(k), \tag{5.32}$$

with

$$\boldsymbol{\nu}(k) = \sum_{j=1}^{N_v(k)} \beta_j(k)\boldsymbol{\nu}_j(k), \tag{5.33}$$

corrects for the measurement uncertainty. Note that Equation (5.30) reduces to the prediction covariance when β_0 equals one.

5.3 Numerical Study

5.3.1 Simulation Environments and Settings

Consider the random-force motion model [41] with process noise covariance $\mathbf{Q}(k) = \mathbf{Q}^*(k) = \mathsf{E}\{\boldsymbol{\omega}(k)\boldsymbol{\omega}^\mathsf{T}(k)\} = \mathbf{I}_2$. Five FTs are located at $(x_1 = 2\text{km}, y_1 = 6\text{km})$, $(x_2 = 12\text{km}, y_2 = 5\text{km})$, $(x_3 = 7\text{km}, y_3 = 12\text{km})$, $(x_4 = 8\text{km}, y_4 = 2\text{km})$ and $(x_5 = 6\text{km}, y_5 = 7\text{km})$. A UE is starting at $\mathbf{x}(0) = [4300\text{m}\ 4300\text{m}\ 15\text{m/s}\ 10\text{m/s}]^\mathsf{T}$. The initial state estimate $\hat{\mathbf{x}}(0|0)$ for the trackers is set as a zero-mean Gaussian random variable with a standard deviation of 50m for the (x,y)-positions and a standard deviation of 4m/s for the velocities around the first true state parameter $\mathbf{x}(0)$ of the trajectory. The initial covariance matrix is set to $\mathbf{P}(0|0) = \mathsf{diag}[50^2\mathrm{m}^2\ 50^2\mathrm{m}^2\ 4^2(\mathrm{m/s})^2\ 4^2(\mathrm{m/s})^2]$. We consider $K = 1000$ time steps and the sampling period is $\Delta t = 0.2\mathrm{s}$.

As in Chapter 4, two different ways of modeling the NLOS occurrences over time are used throughout the simulations. In the first one, NLOS errors are modeled by a two-state Markov chain for each FT and simulation results for the different trackers are given in Section 5.3.2.1. Note that the transition probabilities of the Markov chains used throughout the simulations are given in Appendix A.2.3. In contrast, the second one assumes iid NLOS errors over time and FTs and simulation results can be found in Section 5.3.2.2. For both models a shifted Gaussian and an exponential pdf are used for modeling NLOS errors. Typical values for sensor noise and errors due to NLOS propagation can be found in [92, 111] and we choose $\sigma_G = 150\mathrm{m}$. All simulation results are obtained by averaging over 100 Monte-Carlo runs.

We compare the EKF and the REKF given in Section 4.1.2.1 and 4.1.2.2 for comparison purposes, as in Chapter 4. The residual weighting algorithm from [21, 22], denoted as KF-Rwgh, is used as an other competing tracking algorithm and the proposed approach is denoted as MPDA. Thus, both approaches construct $N = \binom{5}{3} = 10$ subgroups for positioning. For the MPDA, theoretically the independence assumption for the different subgroups from Equation (5.21) is not fulfilled but is still used throughout the simulations. Note that the gate probability for the MPDA tracker is set to $\mathsf{P_G} = 0.99$ corresponding to a false alarm rate of $\mathsf{P_{FA}} = 1 - \mathsf{P_G} = 0.01$.

The standard deviation of the sensor noise is $\sigma_G = 150\mathrm{m}$ and the measurement noise covariance matrices for all trackers are set to $\mathbf{R}^*(k) = \sigma_G^2 \mathbf{I}_M$. A summary of the settings of the different trackers used throughout the simulations can be found in Table 5.1. Apart from the grouping in Section 5.2.2 the MPDA approach has similar computational complexity to an EKF.

Tracker	Parameters	Reference
EKF	-	Section 4.1.2.1
REKF	$c_1 = 1.5$, $c_2 = 3$	Section 4.1.2.2
MPDA	$\mathsf{P_{FA}} = 0.01$	Section 5.2
KF-Rwgh	-	[21, 22]

Table 5.1: Configurations of the different trackers used throughout the simulations

5.3.2 Simulation Results

5.3.2.1 NLOS Outliers Modeled by a Markov Chain

First, the NLOS occurrences for each FT are modeled by a time-homogeneous two-state Markov chain given in Section 2.2.1.2.

Shifted Gaussian Contamination Distribution

We investigate the positioning accuracy of the different trackers when a shifted Gaussian pdf is used for modeling NLOS errors. Figure 5.2 illustrates the MED in m in terms of discrete time index k where 25% of the observations are contaminated by NLOS errors modeled as a shifted Gaussian distribution with mean $\mu_\eta = 1400$m and standard deviation $\sigma_\eta = 400$m.

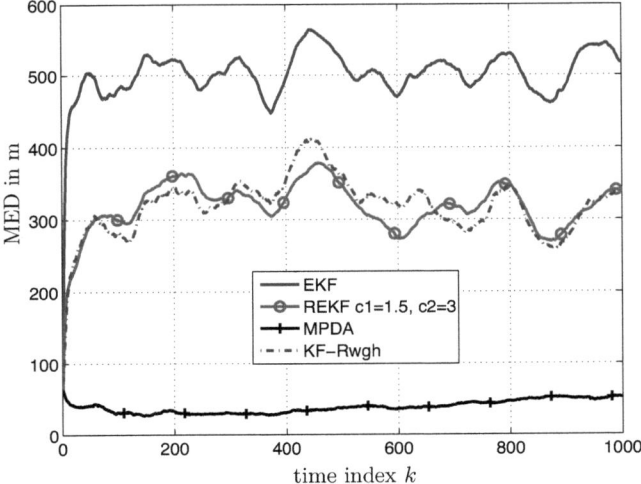

Figure 5.2: MED in m versus time of the different trackers with $\varepsilon = 0.25$ for all FTs. NLOS errors are modeled as a Gaussian pdf with $\mu_\eta = 1400$m and $\sigma_\eta = 400$m and NLOS occurrences at each FT are modeled by a two-state Markov chain.

It can be observed that the EKF suffers from high positioning errors while the EKF and KF-Rwgh gain approximately up to 200m with respect to the EKF. The proposed tracker MPDA significantly outperforms its competitors and achieves an average MED of approximately 40m. Although not shown here, the cdfs of the location errors of the

EKF, REKF and KF-Rwgh slowly converge to one whereas the 99-percentile of the proposed MPDA tracker is at 160m.

Until $\varepsilon = [0.5\ 0.1\ 0.1\ 0.75\ 0.75]$ for the different FTs (corresponds to an average ε of 44%), a significant performance gain of the MPDA algorithm in terms of location error can be observed. However, considering the confidence bounds of the location error, illustrated in Figure 5.3, it can be observed that large position errors with a low frequency occur for the MPDA approach, meaning we can note that the cdf of the location errors slowly converges to one. This is due to the fact that the proposed method does not capture enough information if only two or less LOS measurements are present in consecutive time steps or when the empirical false alarm rate increases, meaning position estimates from LOS FTs are discarded. Then, no valid position estimates are available for the data association and the filter uses the predicted state and covariance as the final estimates leading to large errors or even to loss of the track. Any further increase of ε most often results in divergence of the MPDA filter.

Changing the moments of the Gaussian error pdf leads to similar results.

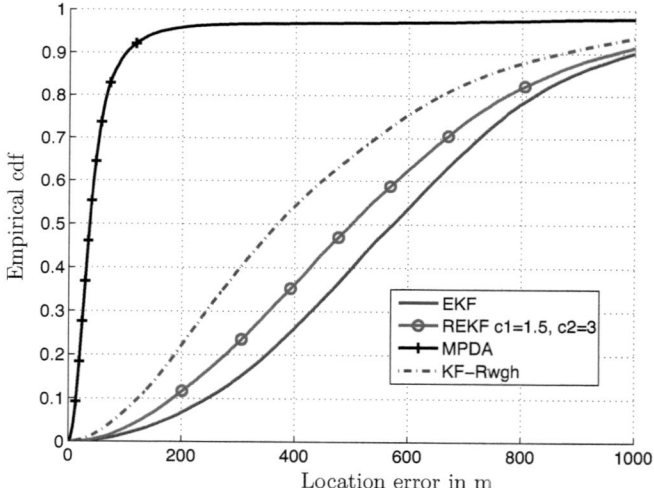

Figure 5.3: Empirical cdf of location errors where $\varepsilon = [0.5\ 0.1\ 0.1\ 0.75\ 0.75]$. NLOS occurrences at each FTs are modeled by a two-state Markov chain and a shifted Gaussian pdf with $\mu_\eta = 1400$m and $\sigma_\eta = 400$m is used for modeling the NLOS errors.

Exponential Contamination Distribution

Next we consider an exponential pdf with $\sigma_\eta = 800$m for modeling the NLOS errors in a scenario where $\varepsilon = 75\%$ of the observations are contaminated by errors due to NLOS propagation. Results in terms of MED versus time are depicted in Figure 5.4.

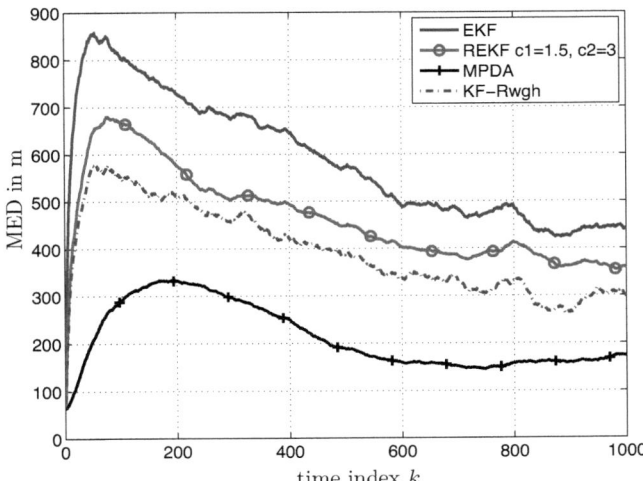

Figure 5.4: MED in m versus time of different trackers with $\varepsilon = 0.75$ for all FTs. NLOS occurrences at each FTs are modeled by a two-state Markov chain with an exponential error pdf with $\sigma_\eta = 800$m.

We observe that the EKF yields highest positioning errors followed by REKF, KF-Rwgh and the proposed MPDA filter which gains on average approximately 130m with respect to the KF-Rwgh. The cdf of the location errors of the different trackers is illustrated in Figure 5.5. We note that the 90-percentile of the MPDA algorithm is at approximately 400m whereas the 90-percentiles of the other trackers widely exceed 600m. While the probability of mis-detection is approximately 3% in the previous examples, where a shifted Gaussian interference pdf is used, it increases here up to 23%. The reason for this is that in general the realizations from the exponential distribution with standard deviation $\sigma_\eta = 800$m are often smaller than those of a shifted Gaussian distribution with $\mu_\eta = 1400$m and $\sigma_\eta = 400$m. This leads to smaller errors in the LS position estimates computed from the different subgroups and consequently to a decrease of the detection probability when an exponential pdf is used. However, as we can observe from the superior results in Figure 5.4 and 5.5, mis-detection of position estimates that are computed from range measurements with smaller noise realizations does not strongly impact the tracking performance since the data association approach in the update step of the Kalman filter partly compensates for them.

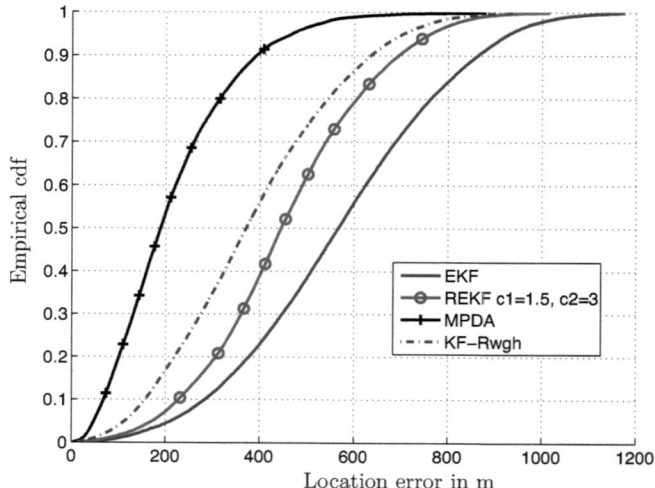

Figure 5.5: Empirical cdf of location errors where $\varepsilon = 0.75$ for all FTs. NLOS occurrences at each FT are modeled by a two-state Markov chain with an exponential error pdf with $\sigma_\eta = 800\text{m}$.

Note that the LOS/NLOS transitions according to a two-state Markov chain is not modeled by any tracker used in the simulations which yields a performance loss in general with respect to the iid case.

Since the proposed MPDA algorithm works more stable over a wider range of ε in an iid environment, we study the behavior of the detector used within this algorithm, given in Section 5.2.3.2, more thoroughly in such an environment.

5.3.2.2 NLOS Outliers Modeled as iid

Shifted Gaussian and Exponential Contamination Distributions
Now we assume that the NLOS error statistics are iid over time and FTs. First, we investigate the probability of detection and false alarm of the proposed MPDA approach in terms of the percentage of NLOS occurrence ε. Table 5.2 illustrates ε, the true and empirical false alarm rates, the probability of NLOS detection P_D and the time average of the MED averaged over 100 Monte-Carlo runs. A shifted Gaussian distribution with mean $\mu_\eta = 1400\text{m}$ and standard deviation $\sigma_\eta = 400\text{m}$ and an exponential distribution with $\sigma_\eta = 800\text{m}$ are used for modeling the NLOS errors. It can be observed for both error statistics that for small to medium ε the empirical

5.3 Numerical Study

false alarm rate obtained through simulations well match the false alarm rate set in (5.14). This is important because we want to exploit the information of all LOS FTs instead of falsely rejecting accurate position estimates. Consequently the MED for both NLOS error statistics is between 24 and 105m. The exact value of one percent, of the nominal false alarm rate is not obtained due to linearization errors included in the Jacobian $\mathbf{H}_n(k)$ and due to the approximation of the test statistics following a chi-squared distribution.

	NLOS error statistics					
	$f_\eta = \mathcal{N}(v; 1400\text{m}, 400^2\text{m}^2)$			$f_\eta = \mathcal{E}(800\text{m})$		
ε	empirical P_{FA}	P_D	E{MED}	empirical P_{FA}	P_D	E{MED}
0	0.012	-	24m	0.012	-	24m
0.1	0.012	0.97	25m	0.014	0.56	25m
0.2	0.02	0.97	28m	0.016	0.58	30m
0.5	0.03	0.98	46m	0.028	0.68	73m
0.6	0.038	0.97	70m	0.044	0.71	105m
0.7	0.325	0.97	483m	0.092	0.75	157m
0.8	0.68	0.97	1392m	0.23	0.78	245m
0.9	0.8	0.97	1850m	0.46	0.83	405m
1	-	0.97	2325m	-	0.87	754m

Table 5.2: Probability of detection and false alarm rates for different NLOS errors modeled as a shifted Gaussian and an exponential pdf. The nominal false alarm rate is set to $P_{FA} = 0.01$.

When we use a shifted Gaussian pdf for modeling NLOS errors high detection rates of 97% are achieved for all ε. This is due to the fact that the LS position estimates computed from NLOS contaminated range measurements possess high errors and consequently strongly differ from the predicted state which simplifies detection of erroneous $\mathbf{z}_n(k)$. However, the limitation of the proposed algorithm becomes apparent when ε increases beyond 60%. Then, for NLOS errors modeled by a shifted Gaussian pdf, the innovation covariance matrix $\mathbf{S}_n(k)$ in (5.10) is underestimated and does not match reality anymore. This results in large values of the test statistic $T_n(k)$ leading to a higher rejection rate for the hypothesis tests in Section 5.2.3.2. Consequently the false alarm rate increases and useful information is discarded leading to divergence of the MPDA filter.

The same argumentation holds for the exponential NLOS error pdf. In this case the probability of detection is lower than P_D for the shifted Gaussian NLOS error pdf because the noise realizations from the exponential pdf $\mathcal{E}(800\text{m})$ are in general smaller which leads to smaller errors in $\mathbf{z}_n(k)$. Then, the difference to the predicted state is less pronounced which makes detection of erroneous $\mathbf{z}_n(k)$ harder. Even though the detection probability of erroneous measurements is low compared to the shifted Gaussian pdf, the proposed tracker yields high precision since the data association step is able to partly mitigate the influence of erroneous position estimates. Again, when ε increases, the innovation covariance matrix $\mathbf{S}_n(k)$ in (5.10) is underestimated leading to larger test statistics $T_n(k)$ and consequently to higher detection rates but also to higher false alarm rates.

Exponential Contamination Distribution

Now we investigate the MED of the different trackers versus the probability of NLOS occurrence ε where the NLOS errors are modeled by an exponential pdf with $\sigma_\eta = 800$m. Figure 5.6 illustrates that the MPDA achieves highest precision for $0 < \varepsilon \leq 0.9$ and gains up to 150m in positioning accuracy with respect to KF-Rwgh. It breaks down and exceeds the MED of the KF-Rwgh and REKF for approximately $\varepsilon > 0.9$. For $0 \leq \varepsilon \leq 0.5$ the REKF and the KF-Rwgh achieve similar positioning accuracy. When ε increases further the REKF loses in precision which is consistent with the theory of *robust statistics* [51].

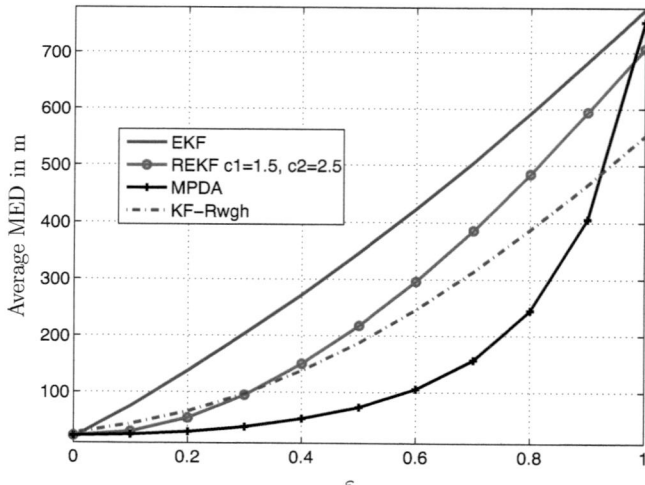

Figure 5.6: Average MED in m versus ε. The contamination pdf $f_\eta = \mathcal{E}(800m)$ and NLOS occurrences are iid.

It is interesting to note that the MED of the EKF increases almost linearly in terms of ε. This is because each element of the measurement noise covariance matrix increases linearly with ε, i.e., $\sigma_V^2 = \sigma_G^2 + \varepsilon\sigma_\eta^2$, leading to a higher impact of the measurements

5.3 Numerical Study

on the state estimates.

Figure 5.7 depicts the MED of the different trackers in terms of the standard deviation of the NLOS errors σ_η, modeled as an exponential pdf where $\varepsilon = 30\%$ of the observations are contaminated by NLOS errors. We observe that the EKF, the REKF and the KF-Rwgh lose in positioning accuracy when σ_η increases since they incorporate all measurements for positioning. In contrast, the MPDA approach discards large outliers and it maintains the same positioning accuracy over a wide range of different σ_η. It is interesting to note that for small σ_η the MED of the MPDA algorithm slightly increases. This is because the NLOS error realizations are less pronounced which make them harder to detect. The consequence is an increase in the mis-detection rate, meaning erroneous position estimates, computed from NLOS FTs, are falsely accepted in the test leading to higher state estimation errors in the update step. In contrast, for large σ_η, it is easier to distinguish position estimates computed from NLOS FTs, which reduces mis-detection rate and only observations from LOS FTs are taken into account. This consequently results in higher precision.

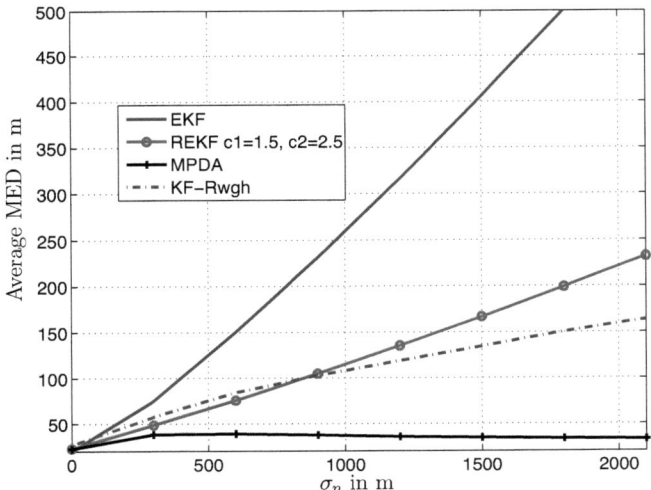

Figure 5.7: Average MED in m versus standard deviation of the NLOS errors σ_η modeled as an exponential pdf with $\varepsilon = 0.3$. NLOS occurrences are iid.

Gaussian Sensor Noise Only

Finally, we study the behavior of the different trackers in the ideal case, meaning when all FTs are in LOS. Figure 5.8 illustrates the MED in terms of the discrete time index

k. As expected the EKF achieves best performance followed by the MPDA and the REKF which achieve similar precision. The KF-Rwgh loses on average 7m with respect to the EKF and 4m with respect to the REKF yielding an average MED of 27m.

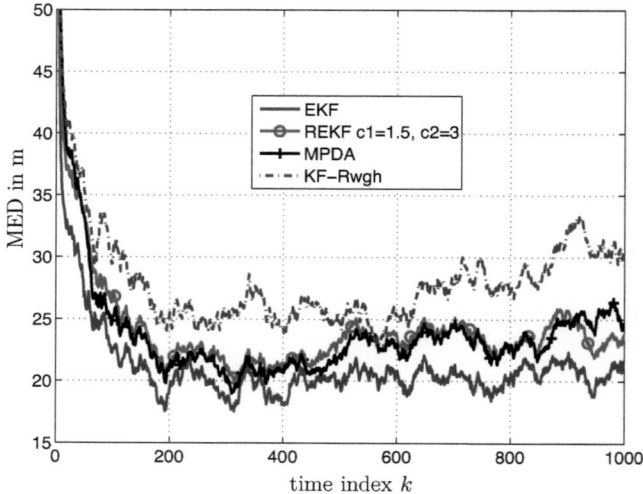

Figure 5.8: MED versus time in LOS environments ($\varepsilon = 0$).

5.4 Discussion and Conclusions

A numerically less complex alternative to the trackers presented in Chapter 4 has been investigated. We propose to construct N subgroups of range measurements to calculate N position estimates via LS. The position estimates with their associated covariance matrices are used in the multiple hypothesis tests to discriminate between LOS/NLOS measurements. Then, the position estimates calculated from erroneous range measurements are discarded. The remaining ones are weighted with different probabilities in a Kalman filter framework.

The proposed approach (MPDA) is compared to the EKF, a robustified EKF based on M-estimation and a competing tracker from the literature denoted as KF-Rwgh. While the complexity of the MPDA approach is slightly higher compared to the complexity of the KF-Rwgh, the former achieves significant improvements with respect to KF-Rwgh, EKF and REKF in NLOS environments whereas similar performance to the EKF is

5.4 Discussion and Conclusions

obtained in environments where all FTs are in LOS.

Since the proposed method does not take into account the NLOS occurrences modeled by a Markov chain it appears to be in general less stable in these situations compared to the iid case. In particular for a shifted Gaussian NLOS error statistics it often diverges when the degree of NLOS contamination exceeds approximately 50%. The reason for this is that the innovation covariance matrix for medium to large ε is underestimated yielding a model mismatch between the innovation sequences and their corresponding covariances. This results in an increase of the false alarm rate in the hypothesis test, meaning a large amount of valid position estimates are rejected. If no position estimates are accepted for consecutive time steps, the MPDA algorithm predicts the state estimates which leads to divergence in most cases.

For the shifted Gaussian NLOS error pdf the detection probability of position estimates computed from NLOS FTs is about 97% because the errors in the LS position estimates are high, which makes them easier to detect. In contrast, even though for an exponential pdf the probability of mis-detection increases, the proposed algorithm remains in general more stable for a wider range of ε because NLOS outliers are smaller. In this case, the data association mitigates the falsely detected position estimates.

Compared to the robust and semi-parametric trackers developed in Chapter 4 the MPDA filter slightly loses in positioning accuracy for small ε. When ε increases the difference becomes more pronounced. For switching LOS/NLOS environments and $\varepsilon > 40\%$ the MPDA filter loses significantly in positioning accuracy compared to the robust and semi-parametric trackers.

Improvements can be achieved by incorporating the jumps of the switching model into the algorithm. Furthermore, estimation of the measurement covariance matrix can increase stability for larger ε since the empirical false alarm rate should then stay closer to the nominal one. This makes it possible to exploit more information contained in the range measurements yielding higher positioning accuracy.

Chapter 6
Conclusions and Future Work

The problem of finding the position of a user equipment (UE) in non-line-of-sight (NLOS) environments based on time-of-arrival (TOA) range estimates has been treated. Interferences due to non-line-of-sight (NLOS) propagation are modeled as outliers which result in a decrease of positioning accuracy when using standard techniques such as least-squares estimation or the extended Kalman filter (EKF).

The two cases of a stationary and a moving UE have been considered separately. A summary of the work performed including conclusions is given in turn for both these situations in Section 6.1. An outlook to future work is given in Section 6.2.

6.1 Conclusions

6.1.1 Stationary User Equipment

For finding the position of a stationary UE, treated in Chapter 2 and Chapter 3, the nonlinear signal model for TOA measurements is reformulated into a linear regression problem. Since conventional least-squares estimators break down when the percentage of NLOS propagation ε increases we propose a semi-parametric positioning algorithm that estimates the interference pdf non-parametrically [43,45]. This estimate is then used for determining the position of the UE based on the maximum likelihood principle. In general, only a small amount of observations are available which makes non-parametric kernel density estimation (KDE) of the noise pdf difficult. This is because outliers produce spurious peaks at the tails of the distribution which leads to convergence problems in the optimization algorithm. To circumvent this problem we propose to use transformation kernel density estimation (TKDE) where the data are transformed by a nonlinear function with a shape parameter which is determined according to the underlying sample such that after transformation the data is *approximately Gaussian*. This allows us to estimate the pdf in the transformed domain with a global bandwidth. A smooth pdf estimate in the original domain is obtained via back-transformation resulting in higher convergence rate for parameter estimation.

Depending on the prior knowledge we have concerning the symmetry of the noise statistics we propose to choose between two different transformation functions [57,112] that are used within the developed semi-parametric estimator [43–45]: The *modulus transformation* [57] decreases the kurtosis of a symmetric sample by transforming the data closer together, an estimator based on this function is described in Section 2.1.4.2 [44]. In contrast, the transformation function from [112], as explained in Section 3.2.3.2, reduces the kurtosis and skewness of an asymmetric sample by transforming large outliers closer to the core of the data while values close to zero are spread further apart resulting in a symmetric, *almost Gaussian* sample. In Chapter 3 the latter transformation is used for positioning based on TOA range measurements [43, 45] because the NLOS errors are assumed to be positive resulting in skew-symmetric observations. The degree of asymmetry of the sample is reflected in the automatically selected shape parameter for the transformation which is used to transform the residuals and obtain the required pdf estimate.

The proposed estimator achieves positioning accuracy similar to least-squares in LOS environments and allows for a significant gain in performance with respect to conventional and robust M-estimators when the percentage of errors due to NLOS effects increases. In particular for skew-symmetric NLOS error statistics with an exponential, lognormal or Rayleigh pdf the proposed estimator achieves large performance gains with respect to conventional and robust methods over wide ranges of ε. For shifted Gaussian NLOS error statistics and large ε the gain in positioning accuracy is less pronounced.

6.1.2 Moving User Equipment

Incorporating a motion model for a moving UE decreases positioning errors with respect to the stationary case because of time averaging effects. In this context optimal solutions briefly introduced in Chapter 2 are intractable. Instead suboptimal algorithms such as the EKF achieve high precision in LOS environments. However, positioning errors increase severely with the percentage of NLOS errors.

In Chapter 4 we adapt the semi-parametric estimator developed in Chapter 3 to the problem of a moving UE, denoted as EKF-SP-MR. The motivation in doing so is to avoid the trade-off between efficiency and robustness which is inherent for classical robust estimators. For this purpose, the EKF equations are reformulated as a linear regression problem and solved at each time step using the semi-parametric estimation scheme. Since it is not straightforward to establish the asymptotic covariance of the semi-parametric estimator the posterior covariance of the standard EKF is used instead as an approximation.

A parametric alternative [47], denoted as R-IMM, is also developed in Chapter 4 where the EKF equations again rewritten as linear regression problem so that robust estimation techniques can be applied. In general, robust estimators always trade-off efficiency in the nominal cases versus robustness in the NLOS case so that achieving both with a single estimator is not possible. To overcome this problem an EKF, well suited for LOS channels, is run in parallel with a robust EKF (REKF) based on robust regression, well suited for NLOS environments. The second filter is adjusted so that it clips a large amount of the data and consequently behaves robustly in highly contaminated NLOS environments. The likelihood of each filter matching the underlying situation is calculated and incorporated in the final position estimate which is a weighted combination of the state estimates computed by both filters. In LOS environments, the weights of the EKF tend to one, whereas the weights of the REKF tend to zero, meaning the influence of the EKF on the state estimates is increased whereas the influence of the REKF is reduced. In severe NLOS environments the opposite is true.

The R-IMM and the EKF-SP-MR trackers both significantly outperform conventional and robust techniques in NLOS environments and gain up to 200m in terms of positioning accuracy. In particular for non-Gaussian error statistics which follow an exponential or Rayleigh distribution, the EKF-SP-MR tracker gains in precision with respect to R-IMM because the higher order moments of the noise pdf are incorporated into the estimation procedure via non-parametric pdf estimation.

In contrast, the R-IMM achieves higher precision when the NLOS errors follow a shifted Gaussian pdf since it discards larger outliers whereas the semi-parametric tracker approximates the noise pdf, which becomes difficult when the pdf is multimodal.

While the R-IMM approach achieves performance similar to the EKF in LOS environments, accuracy of the semi-parametric tracker drops by a few meters in this situation. This can be explained by the fact that the latter possesses more degrees of freedom and consequently more uncertainty in the nominal case, as it relies on non-parametric pdf estimation. Furthermore, unlike the EKF-SP-MR, the R-IMM algorithm takes into account the switching of the LOS/NLOS events that are modeled by a two-state Markov chain for each FT and also contributes to the performance gain.

The main limitation of the R-IMM, in particular the REKF used within the R-IMM, is that its precision significantly decreases for large percentages of NLOS occurrences and large magnitudes of NLOS measurements because it is based on robust regression. According to the theory of *robust statistics* a robust M-estimator breaks down when the degree of outliers achieves 50%. However, since a motion model is incorporated within a Bayesian framework the proposed R-IMM algorithm exhibits numerically stable performances for up to 60% outliers. Increasing ε further leads to high positioning errors given that the magnitude of the NLOS errors is large compared to the sensor noise.

To summarize, the EKF-SP-MR is preferable in environments where the NLOS interference statistics are completely unknown and when a large number of FTs are available

6.1 Conclusions

because it relies on non-parametric KDE. However, high positioning accuracy is obtained in the simulations for only five FTs. In contrast, the R-IMM is better suited if the NLOS errors statistics are likely to follow a shifted Gaussian distribution, for small to medium ε and also for smaller sample sizes since it is based on parametric robust M-estimation.

While both trackers are roughly comparable in terms of computational complexity, a numerically simpler tracking scheme is developed in Chapter 5.

The MPDA approach presented in Chapter 5 relies on a joint error detection and tracking scheme and is published in [48]. We propose to construct different subgroups of TOA range measurements together with the corresponding positions of the FTs. Each subgroup provides a least-squares position estimate of the UE together with its covariance matrix that are both used in a hypothesis test for NLOS detection. The position estimates computed by NLOS FTs are discarded whereas the accepted measurements are weighted with different probabilities in a Kalman filter framework. Again, significant improvements can be achieved with respect to conventional and robust techniques in NLOS environments whereas similar performance to the EKF is obtained in LOS environments.

The limitations of this tracker become apparent when the percentage of NLOS outliers is close to 50% and the magnitude of the NLOS outliers is large, such as when they are modeled by a shifted Gaussian pdf. In such cases, the innovation covariance matrix calculated in the Kalman filter recursions does not match the observed innovation sequences which lead to higher false alarm rates. Thus, useful information is lost because valid position estimates are discarded. If this happens for consecutive time step, the state estimates for these time steps are only based on prediction which lead most often to divergence and loss of the track.

For small ε up to approximately 30% and iid environments the MPDA tracker proposed in Chapter 5 has slightly less positioning accuracy than R-IMM and EKF-SP-MR. However, the difference becomes more pronounced for larger ε and when the LOS/NLOS occurrences are modeled by a Markov chain. In these environments, it is less stable than R-IMM and EKF-SP-MR and produces large positioning errors. For $\varepsilon > 60\%$ and exponentially distributed NLOS outliers, the semi-parametric tracker outperforms the R-IMM and MPDA algorithms significantly.

6.2 Future Work

6.2.1 Stationary User Equipment

For TOA positioning, the MLE for the shape parameter of the transformation function in [112] can be replaced by a more robust estimation scheme such as that in [16], in order to reduce the influence of large outliers in the sample. This will emphasize the impact of the core of the data on the pdf estimate and consequently can result in higher accuracy in different environments.

When other signal parameters such as AOA, RSS, TDOA or a combination thereof are available it becomes desirable to adapt the semi-parametric estimation scheme to these measurements. In this context, linearization of the nonlinear relationships between the signal parameters and the position of the UE can result in stochastic elements and even outliers in the regressor matrix, also known as *leverage points*. These perturbations result in higher estimation errors when using classical, robust and the proposed semi-parametric estimator because they all assume that the regressor matrix is deterministic and consequently do not mitigate outliers in the parametric model.

To protect against errors in the regressor matrix generalized M-estimators [68] can be deployed. These techniques reduce the impact of *leverage points* on the parameter estimate leading to superior results compared to conventional robust schemes. Combining generalized M-estimators with the semi-parametric approach results in down-weighting of large *leverage points*. Thus, incorporating this weighting scheme allows for adaptivity to the noise pdf and robustness against deviations from the presumed model.

6.2.2 Moving User Equipment

To increase positioning accuracy for a moving UE estimation of the posterior covariance matrix of the semi-parametric tracker can be improved. Instead of using the posterior covariance from a conventional EKF an estimate of the covariance of the semi-parametric estimator can be deployed. In particular for small samples this is a difficult task and resampling techniques [114] can be used for this purpose. For larger sample sizes we can also use the asymptotic result from Equation (2.13) which defines the asymptotic covariance matrix of any estimator in terms of its score function and the underlying parametric model. To do so, the parametric score function is replaced by its non-parametric estimate and the integrals by sums in order to obtain an estimate of the covariance for finite samples.

Further improvements of the semi-parametric (noise-adaptive) tracker can be achieved by incorporating the switching of LOS/NLOS occurrences into the tracking scheme. To do so, we can take into account information on the shape parameter of the parametric transformation function from the previous time step into the estimation of this quantity at the actual time step. For consecutive time instants the shape parameter then remains similar which better reflects the underlying LOS/NLOS situation and should consequently contribute to the stability of TKDE. An increase in positioning accuracy is then expected.

While the semi-parametric tracker implicitly estimates the measurement noise covariance via non-parametric pdf estimation, improvements for the R-IMM can be achieved if this quantity is estimated directly from the measurements. This makes the approach less sensitive to observations with large noise variances.

It is straightforward to extend the R-IMM algorithm to other signal parameters such as TDOA and RSS or a combination thereof. Depending on the prior knowledge we have about the NLOS error statistics the soft-limiter or a redescending score function can be used for the REKF in the R-IMM algorithm.

In contrast, using a combination of different signal parameters for the semi-parametric approach seems difficult because different signal parameters can have different NLOS error characteristics. Thus, the choice of a transformation function for TKDE and its corresponding parameter is unclear.

For the MPDA algorithm other signal parameters can be incorporated because the proposed algorithm first calculates various position estimates by exploiting the redundancy of the measurements and then uses the computed estimates in a detection and tracking framework. Consequently, using additional signal parameters should increase position accuracy.

When the degree of NLOS contamination exceeds $40\% - 50\%$ the false alarm rate in the NLOS detector of the MPDA algorithm and consequently the positioning errors increase. This is due to the mismatch between the observed innovation sequence and its estimated covariance.

Improvements can be achieved by estimating the measurement covariance at each time step. This quantity can be used to determine the innovation covariance more accurately resulting in a more stable detector for different percentages of NLOS contamination. Consequently higher positioning accuracy can be achieved by exploiting a larger amount of the valid position estimates.

A.1　Choice of Clipping Parameters for Geolocation

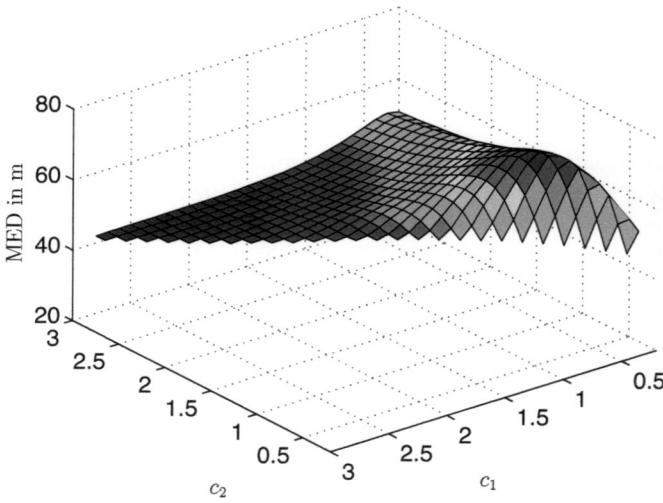

Figure A.1: MED of the redescending score function [24] versus different clipping parameters c_1 and c_2 in LOS ($\varepsilon = 0$) with $\sigma_G = 150$m for $K = 5$ time steps, meaning $KM = 50$ observations (see Figure 3.4).

A.1 Choice of Clipping Parameters for Geolocation

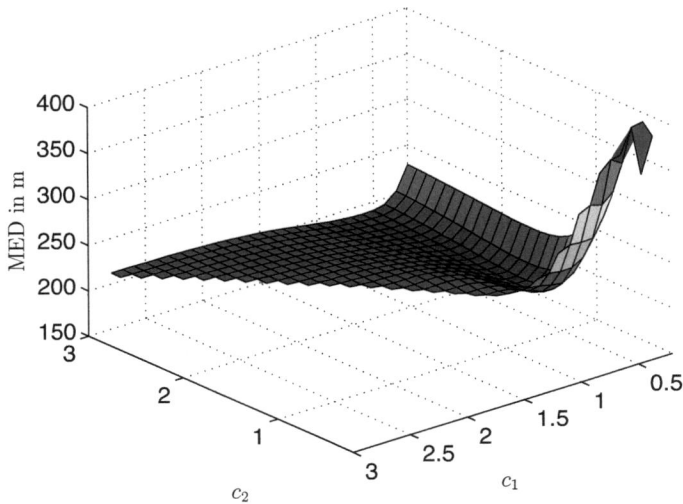

Figure A.2: MED of the redescending score function [24] versus different clipping parameters c_1 and c_2 in NLOS ($\varepsilon = 0.4$) with $\sigma_G = 150$m, $\sigma_\eta = 300$m and $\mu_\eta = 1000$m for $K = 5$ time steps, meaning $KM = 50$ observations (see Figure 3.4).

A.2 Robust Tracking

A.2.1 Choice of Clipping Parameters

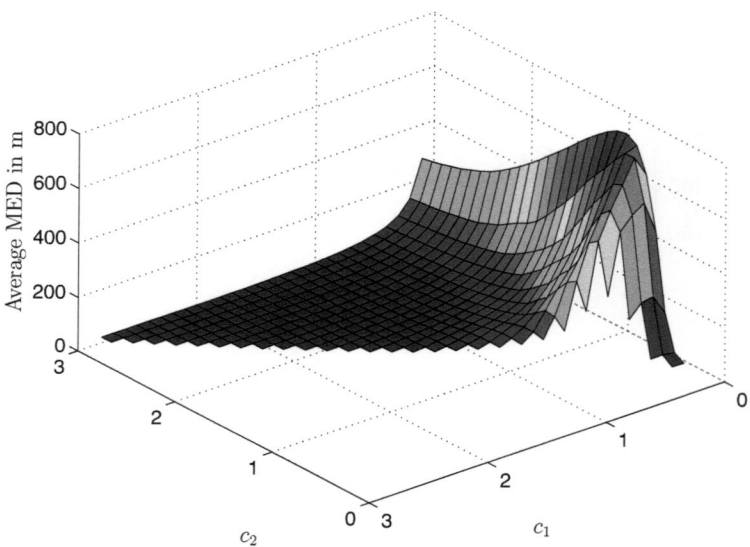

Figure A.3: Average MED of the redescending score function [24] versus different clipping parameters c_1 and c_2 in LOS ($\varepsilon = 0$) with $\sigma_G = 150$m for $M = 5$ FTs from Figure 4.2.

A.2 Robust Tracking

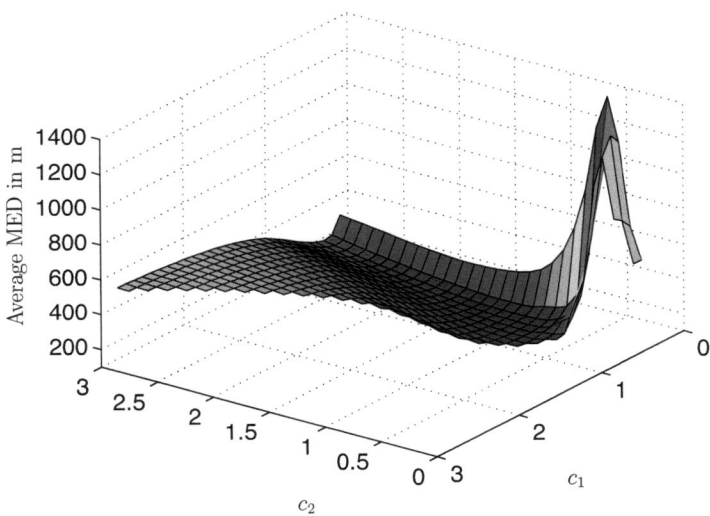

Figure A.4: Average MED of the redescending score function [24] for different clipping parameters c_1 and c_2 in NLOS ($\varepsilon = 0.4$) with $\sigma_G = 150$m, $\sigma_\eta = 300$m and $\mu_\eta = 1000$m for $M = 5$ FTs from Figure 4.2. NLOS occurrences are modeled as iid.

A.2.2 Transition Probabilities of Reduced Model

For the reduced model in Table 4.3 we can calculate the following transition probabilities. Let

$$\Pr(\mathcal{M}_j) = \lim_{k \to \infty} \Pr\{\mathcal{M}(k) = \mathcal{M}_j\}.$$

The probability to change from the nominal case where all FTs are in LOS to any other model is p'_{12} given as

$$p'_{12} = \Pr\{\mathcal{M}_2 \cup \mathcal{M}_3 \cup \ldots \cup \mathcal{M}_r | \mathcal{M}_1\} = \sum_{j=2}^{r} \Pr\{\mathcal{M}_j(k) | \mathcal{M}_1(k-1)\} = \sum_{j=2}^{r} p_{1j} \quad (\text{A.1})$$

In contrast, the transition probability to switch from any model where at least one FT is affected by NLOS propagation to the nominal case is p'_{21} and given as

$$p'_{21} = \Pr\{\mathcal{M}_1 | \mathcal{M}_2 \cup \ldots \cup \mathcal{M}_r\} = \frac{\Pr\{\mathcal{M}_2 \cup \mathcal{M}_3 \cup \ldots \cup \mathcal{M}_r | \mathcal{M}_1\} \Pr\{\mathcal{M}_1\}}{\Pr\{\mathcal{M}_2 \cup \mathcal{M}_3 \cup \ldots \cup \mathcal{M}_r\}} \quad (A.2)$$

$$= \frac{\sum_{j=2}^{r} \Pr\{\mathcal{M}_j(k) | \mathcal{M}_1(k-1)\} \Pr\{\mathcal{M}_1\}}{\sum_{j=2}^{r} \Pr\{\mathcal{M}_j\}} \quad (A.3)$$

$$= \frac{\sum_{j=2}^{r} p_{1j} \Pr\{\mathcal{M}_1\}}{\sum_{j=2}^{r} \Pr\{\mathcal{M}_j\}}, \quad (A.4)$$

where we have used Bayes Theorem and the fact that the events $\mathcal{M}_j$, $j = 1, 2, \ldots, r$ are mutually exclusive.

A.2.3 Markov Matrices

Recall Section 2.2.1.2. A first order two-state time-homogeneous Markov chain has transition matrix $\mathbf{\Pi}$ defined as [25]

$$\mathbf{\Pi} = \begin{bmatrix} 1 - \pi_{12} & \pi_{12} \\ \pi_{21} & 1 - \pi_{21} \end{bmatrix}, \quad (A.5)$$

where π_{12} is the transition probability from state LOS to NLOS and π_{21} is the transition probability from state NLOS to LOS.

Table A.1 summarizes the values for π_{12} and π_{21} we use for the simulations in Chapter 4 and 5 for particular values of the probability of NLOS occurrence ε, confer (2.36).

Table A.1: Parameters for Markov transition matrices (A.5) for modeling different probabilities of NLOS occurrence for a given BS.

ε	π_{12}	π_{21}
0.1	0.01	0.09
0.25	0.006	0.02
0.5	0.02	0.02
0.75	0.06	0.02

List of Acronyms

AKDE	adaptive kernel density estimation
AOA	angle-of-arrival
cdf	cumulative distribution function
CLT	Central Limit Theorem
CRLB	Cramér-Rao Lower Bound
EKF	extended Kalman filter
FT	fixed terminal
GNSS	Global Navigation Satellite System
GPB	generalized pseudo-Bayesian
GPS	Global Positioning System
GSM	Global System for Mobile Communications
iid	independent and identically distributed
IMM	interacting multiple model
KDE	kernel density estimation
KF	Kalman filter
LOS	line-of-sight
LS	least-squares
mad	median or mean absolute deviation
MAP	maximum a posteriori
MC	Markov chain
MED	mean error distance
ML	maximum likelihood
MLE	maximum likelihood estimator
MMSE	minimum mean-square error

MPDA	modified probabilistic data association
MSE	mean square error
MUD	multiuser detection
NLOS	non-line-of-sight
PCRLB	Posterior Cramér-Rao Lower Bound
PDA	probabilistic data association
pdf	probability density function
PF	particle filter
REKF	robust extended Kalman filter
RMSE	root mean square error
RSS	received-signal-strength
RTOF	round trip time of flight
RTT	round trip time
rv	random variable
SNR	signal-to-noise-ratio
TA	timing advance
TDOA	time-difference-of-arrival
TKDE	transformation kernel density estimation
TOA	time-of-arrival
TPM	transition probability matrix
UE	user equipment
UKF	unscented Kalman filter
UMTS	Universal Mobile Telecommunications System
WLS	weighted least-squares

List of Symbols

$\mathbf{a}(\cdot)$	Non-linear function describing state transition
$\mathbf{A}$	State transition matrix
b	Smoothing parameter for redescending score function
$\mathbf{B}$	Measurement matrix in the linearized model
c	Normalization constant
c_1	Clipping point for Huber's soft-limiter, and first clipping point of redescending score function
c_2	2nd clipping point of redescending score function
$\mathbf{C}(k)$	Factorization matrix obtained through Cholesky decomposition
$\mathbf{D}$	Regressor matrix for the linear model
$\mathsf{E}\{\cdot\}$	Expectation operator
$\mathcal{E}(\sigma)$	Exponential pdf with standard deviation σ
E_j	Event that the j-th position estimate is computed by LOS FTs
E_0	None of the position estimates stem from LOS FTs
$\mathcal{F}$	Class of mixture distributions
$f_V(v)$	Probability density function of noise and interferences
$f'_V(v)$	Derivative of $f_V(v)$
$f_{\tilde{V}}(\tilde{v})$	Probability density function of noise and interferences after linearization
$f'_{\tilde{V}}(\tilde{v})$	Derivative of $f_{\tilde{V}}(\tilde{v})$
$f_W(w)$	Transformed probability density function
f_η	pdf of NLOS error statistics
$f_{\chi^2(2)}(x)$	Chi-squared pdf with two degrees of freedom
$f(\mathbf{y}(k)\|\mathbf{x}(k))$	Measurement noise pdf
$f(\mathbf{x}(k)\|\mathbf{x}(k-1))$	Process noise pdf
$f(\mathbf{x}(k)\|\mathbf{Y}^{k-1})$	Prediction pdf
$f(\mathbf{x}(k)\|\mathbf{Y}^k)$	Posterior pdf
$g_m(k)$	Realization of a Gaussian random variable for the m-th FT at time step k
$\mathbf{G}$	Mapping matrix from acceleration noise to speed and velocity
$\mathbf{h}(\cdot)$	Nonlinear measurement model
$h_m(\cdot)$	m-th element from $\mathbf{h}(\cdot)$
$\mathbf{H}(k)$	Jacobian matrix of $\mathbf{h}(\cdot)$
$\mathcal{H}(v)$	Contamination distribution in mixture model

H_0	Null hypothesis
H_1	Alternative hypothesis
$\mathbf{I}_M$	$M \times M$ identity matrix
J	Number of basis functions
$\mathbf{J}(\boldsymbol{\theta})$	Fisher information matrix
k	Discrete time index
K	Number of time steps
$\mathcal{K}(\cdot)$	Kernel for KDE
$\mathbf{K}(k)$	Kalman gain
l	Iteration index
$L(\boldsymbol{\theta}\|\mathbf{y}(k))$	Likelihood function
m	Index for FTs
M	Number of FTs
$\mathcal{M}_j$	$j = 1, \ldots, r$, mode variable of complete model
$\mathcal{M}'_j$	$j = 1, 2$, mode variable of reduced model
$\mathcal{M}(k)$	Mode in effect at time step k
n	Subgroup index for MPDA approach
N	Number of subgroups
N_v	Number of validated subgroups
N_x	Dimension of the state vector $\mathbf{x}(k)$
$\mathcal{N}(x, \mu, \sigma^2)$	Gaussian pdf with mean μ and variance σ^2
o	Step parameter for Newton-Raphson algorithm
p_{ij}	Transition probability from mode $\mathcal{M}_i$ to $\mathcal{M}_j$
p'_{ij}	Transition probability from mode $\mathcal{M}_i$ to $\mathcal{M}_j$ for the reduced model
$\mathbf{P}(k\|k-1)$	Prediction covariance matrix
$\mathbf{P}(k\|k)$	Posterior covariance matrix
$\mathbf{P}_j(k\|k)$	Posterior covariance matrix of the j-th filter in the IMM algorithm
$\mathbf{P}_{0j}(k-1\|k-1)$	Initialization covariance matrix of the j-th filter in the IMM algorithm
P_D	Empirical probability of detection
P_G	Gate probability
P_{FA}	Nominal false alarm rate
$\mathbf{Q}(k)$	True process noise covariance matrix
$\mathbf{Q}^*(k)$	Selected process noise covariance matrix
r	Number of different modes for the jump-nonlinear model
$\mathbf{R}(k)$	True measurement noise covariance matrix

$\mathbf{R}^*(k)$	Selected measurement noise covariance matrix
$\mathbf{R}_1^*(k)$	Selected measurement noise covariance matrix of EKF in R-IMM
$\mathbf{R}_2^*(k)$	Selected measurement noise covariance matrix of REKF in R-IMM
s	Random sequence of LOS/NLOS occurrences
$\mathbf{S}(k)$	Innovation covariance matrix
$t(v, \lambda)$	Transformation function for symmetrizing a sample
$\mathbf{T}$	Markov matrix of the augmented model
$\mathbf{T}'$	Markov matrix of the reduced model
$\boldsymbol{u}$	Weights for adaptive parametric estimation
$v(k)$	Noise component before linearization
$\hat{v}(k)$	Estimate of $v(k)$
$\tilde{v}(k)$	Noise component after linearization
$\hat{\tilde{v}}(k)$	Estimate of $\tilde{v}(k)$
$\mathbf{v}(k)$	Noise vector before linearization
$\hat{\mathbf{v}}(k)$	Vector of residuals
$\tilde{\mathbf{v}}(k)$	Noise vector after linearization
$\hat{\tilde{\mathbf{v}}}(k)$	Vector of residuals after linearization
V	Random variable describing the mixture noise
$\mathcal{V}_j(k)$	$j = 1, \ldots, N_v$, surface of validation region
$\mathbf{W}$	Weighting matrix used within WLS estimator
$\mathbf{W}^*$	Optimal weighting matrix for WLS estimator
W	Random variable describing transformed residuals
$\mathbf{w}$	Vector of transformed residuals
$\mathbf{w}_s$	Symmetrized vector of transformed residuals
$\mathbf{x}(k)$	State vector
$\hat{\mathbf{x}}_j(k\|k)$	Updated state vector for the j-th filter in the IMM algorithm
$\hat{\mathbf{x}}_{0j}(k\|k)$	Initial state vector for the j-th filter in the IMM algorithm
$x(k)$	x-position of the UE at time step k
$\dot{x}(k)$	Velocity in x-direction of the UE at time step k
$x_{\mathsf{FT},m}$	x-coordinate of the m-th FT
$\hat{\mathbf{x}}(k\|k)$	Posterior state vector
$\hat{\mathbf{x}}(k\|k-1)$	Predicted state vector
$y(k)$	y-position of the UE at time step k
$\dot{y}(k)$	Velocity in y-direction of the UE at time step k
$y_{\mathsf{FT},m}$	y-coordinate of the m-th FT

$\mathbf{y}(k)$	observations
$\mathbf{Y}^k$	Sequence of $\mathbf{y}(k)$, $k = 1, \ldots, K$
$\mathbf{z}_n(k)$	Least-squares position estimate $n = 1, \ldots, N$
$\hat{\mathbf{z}}(k\|k-1)$	Predicted position
$\mathbf{Z}^k$	Sequence of $\mathbf{z}(k)$, $k = 1, \ldots, K$
α	Break condition for Newton-Raphson algorithm
β_j	Association probabilities of the validated measurements
γ	Threshold for NLOS detection
δ	Global bandwidth for KDE
Δt	Sampling time
ε	Degree of NLOS contamination
ζ	Shape parameter of ad-hoc transformation function
$\eta_m(k)$	Realization of the contamination distribution f_η for the m-th FT at time step k
$\boldsymbol{\theta}$	Vector parameter of interest $\boldsymbol{\theta} = [x\,y]^\mathsf{T}$
κ	Time index for a specific time step
λ	Shape parameter of transformation function
$\Lambda_j(k)$	Likelihood function in R-IMM algorithm of the j-th filter
$\mu_{i\|j}(k-1\|k-1)$	Mixing probability in IMM algorithm
$\mu_j(k)$	Prior probability of the j-th filter in the IMM algorithm
μ_η	Mean value of contamination distribution
μ_W	Mean value of transformed residuals
$\boldsymbol{\nu}(k)$	Innovation sequence in EKF
$\boldsymbol{\xi}(k)$	Noise vector of Gauss-Newton algorithm
π_{ij}	Transition probabilities of two-state Markov chain
$\mathbf{\Pi}$	Transition probability matrix of Markov chain
$\rho(v)$	Penalty function
$\rho_{c_1}(v)$	Penalty function for soft-limiter
σ_G	Standard deviation of sensor noise
σ_η	Standard deviation of NLOS errors
σ_V	Standard deviation of mixture pdf
$\sigma_{\tilde{V}}$	Standard deviation of mixture pdf after linearization
σ_W	Standard deviation of transformed residuals
$\Phi(v)$	Standard Gaussian cdf evaluated at v
$\boldsymbol{\phi}(v)$	Vector of basis functions for adaptive parametric estimation
$\phi(v)$	Basis function for adaptive parametric estimation

ϑ	Threshold in definition of consistency
$\varphi(v)$	Location score function
$\psi(v)$	Robust location score function
$\psi_{c_1}(v)$	Huber's soft-limiter
$\psi_{c_1,c_2}(v)$	Redescending score function
$\boldsymbol{\omega}(k)$	Process noise vector in state space model
$\boldsymbol{\Omega}$	Weighting Matrix in robust and semi-parametric algorithms
$\text{Cov}\{\cdot\}$	Covariance
$\dim(\cdot)$	Dimension of a vector
$\exp(\cdot)$	Exponential function
$\log(\cdot)$	Natural logarithm
$\max(\cdot)$	Maximum of a vector
$\text{sign}(\cdot)$	Signum function
$\text{var}(\cdot)$	Variance
$*$	Convolution operator
$\lvert \cdot \rvert$	Absolute value for scalars and vectors and determinant for matrices
$\lVert \cdot \rVert$	L1-norm
T	Transpose of a vector or matrix

Bibliography

[1] B. D. Anderson and J. B. Moore. *Optimal Filtering*. Englewood-Cliffs, NJ: Prentice-Hall, 1979.

[2] I. Arasaratnam and S. Haykin. Cubature Kalman filters. *IEEE Transactions on Automatic Control*, 54(6):1254–1269, 2009.

[3] M. S. Arulampalam, S. Maskell, N. Gordon, and T. Clapp. A tutorial on particle filters for online nonlinear/non-Gaussian Bayesian tracking. *IEEE Transactions on Signal Processing*, 50(2):174–188, 2002.

[4] G. Audrey and G. Julie. An unscented Kalman filter based maximum likelihood ratio for NLOS bias detection in UMTS localization. In *16th European Signal Processing Conference (EUSIPCO)*, Lausanne, Switzerland, August 2008.

[5] Y. Bar-Shalom, X. Rong Li, and T. Kirubarajan. *Estimation with Applications to Tracking and Navigation*. John Wiley & Sons, 2001.

[6] R. Beran. Adaptive estimates for autoregressive processes. *Annals of the Institute of Statistical Mathematics*, 28(1):77–89, 1976.

[7] A. Berlinet and L. Devroye. A comparison of Kernel Density Estimates. *Publications de l'Institut de Statistique de l'Université de Paris*, 38(3):3–59, 1994.

[8] P. J. Bickel, C. A. J. Klaassen, Y. R. Ritov, and J. A. Wellner. *Efficient and Adaptive Estimation for Semiparametric Models*. Springer, 1993.

[9] C. G. Boncelet and B. W. Dickinson. An approach to robust Kalman filtering. In *The 22nd IEEE Conference on Decision and Control*, pages 304–305, 1983.

[10] J. Borras, P. Hatrack, and N. B. Mandayam. Decision theoretic framework for NLOS identification. In *48th IEEE Vehicular Technology Conference*, volume 2, pages 1583–1587, 1998.

[11] S. Boyd and L. Vandenberghe. *Convex Optimization*. Cambridge University Press, 2009.

[12] R. F. Brcich and A. M. Zoubir. Robust estimation with parametric score function estimation. In *IEEE International Conference on Acoustics, Speech, and Signal Processing, (ICASSP '02)*, volume 2, pages 1149–1152, 2002.

[13] J. Caffery. *Wireless location in CDMA cellular radio systems*. Kluwer Academic Publishers, 1999.

[14] J. Caffery and G. L. Stuber. Overview of radiolocation in CDMA cellular systems. *IEEE Communications Magazine*, 36(4):38–45, 1998.

[15] J. Caffery and G. L. Stuber. Subscriber location in CDMA cellular networks. *IEEE Transactions on Vehicular Technology*, 47(2):406–416, 1998.

[16] R. J. Carroll. A robust method for testing transformation to achieve approximate normality. *Journal of the Royal Statistical Society, Series B*, 42:71–78, 1980.

[17] G. C. Carter. Coherence and time delay estimation. *Proceedings of the IEEE*, 75(2):236–255, 1987.

[18] R. Casas, A. Marco, J. J. Guerrero, and J. Falcó. Robust estimator for non-line-of-sight error mitigation in indoor localization. *EURASIP J. Appl. Signal Process.*, 1(1):156–156, 2006.

[19] Y. T. Chan and K. C. Ho. A simple and efficient estimator for hyperbolic location. *IEEE Transactions on Signal Processing*, 42(8):1905–1915, 1994.

[20] B.-S. Chen, C.-Y. Yang, F.-K. Liao, and J.-F. Liao. Mobile location estimator in a rough wireless environment using extended Kalman-based IMM and data fusion. *IEEE Transactions on Vehicular Technology*, 58(3):1157–1169, 2009.

[21] P.-C. Chen. A cellular based mobile location tracking system. In *IEEE 49th Vehicular Technology Conference*, volume 3, pages 1979–1983, 1999.

[22] P.-C. Chen. A non-line-of-sight error mitigation algorithm in location estimation. In *IEEE Wireless Communications and Networking Conference, (WCNC'99)*, pages 316–320 vol.1, 1999.

[23] K. W. Cheung, H. C. So, W.-K. Ma, and Y. T. Chan. Least squares algorithms for time-of-arrival-based mobile location. *IEEE Transactions on Signal Processing*, 52(4):1121–1130, 2004.

[24] J. R. Collins. Robust estimation of a location parameter in the presence of asymmetry. *Annals of Statistics*, 4:68–85, 1976.

[25] T. M. Cover and J. A. Thomas. *Elements of Information Theory*. John Wiley and Sons, 1991.

[26] L. Devroye and G. Lugosi. Variable kernel estimates: On the impossibility of tuning the parameters. In *High-Dimensional Probability II*, pages 405–424. Springer-Verlag, 1998.

[27] L. P. Devroye and G. Lugosi. *Combinatorial methods in density estimation*. Springer, 2001.

[28] Z. M. Durovic and B. D. Kovacevic. Robust estimation with unknown noise statistics. *IEEE Transactions on Automatic Control*, 44(6):1292–1296, 1999.

[29] F. El-Hawary and Y. Jing. Robust regression-based EKF for tracking underwater targets. *IEEE Journal of Oceanic Engineering*, 20(1):31–41, 1995.

[30] S.-H. Fang, T.-N. Lin, and K.-C. Lee. A novel algorithm for multipath fingerprinting in indoor WLAN environments. *IEEE Transactions on Wireless Communications*, 7(9):3579–3588, 2008.

[31] W. H. Foy. Position-location solutions by taylor-series estimation. *IEEE Transactions on Aerospace and Electronic Systems*, AES-12(2):187–194, 1976.

[32] J. Friedmann, H. Messer, and J.-F. Cardoso. Robust parameter estimation of a deterministic signal in impulsive noise. *IEEE Transactions on Signal Processing*, 48(4):935–42, April 2000.

[33] C. Fritsche, U. Hammes, A. Klein, and A. M. Zoubir. Robust mobile terminal tracking in NLOS environments using interacting multiple model algorithm. In *IEEE International Conference on Acoustics, Speech, and Signal Processing (ICASSP '09)*, Taipei, Taiwan, April 2009.

[34] C. Fritsche and A. Klein. Nonlinear filterung for hybrid GPS/GSM mobile terminal tracking. *International Journal of Navigation and Observation*, 2009. submitted to.

[35] S. Gezici. A survey on wireless position estimation. *Wireless Personal Communications*, 44(3):263–282, February 2008.

[36] S. Gezici, H. Kobayashi, and H. V. Poor. Nonparametric non-line-of-sight identification. In *IEEE 58th Vehicular Technology Conference*, volume 4, pages 2544–2548, 2003.

[37] S. Gezici and H. V. Poor. Position estimation via ultra-wide-band signals. *Proceedings of the IEEE*, 97(2):386–403, 2009.

[38] S.J. Godsill and P.J. Rayner. *Digital Audio Restoration*. 1998.

[39] G. Golub and C. Van Loan. *Matrix Computation*. Johns Hopkins Studies in the Mathematical Sciences, 1996.

[40] E. Grosicki and K. Abed-Meraim. A new trilateration method to mitigate the impact of some non-line-of-sight errors in TOA measurements for mobile localization. In *IEEE International Conference on Acoustics, Speech, and Signal Processing (ICASSP '05)*, volume 4, pages 1045–1048, 2005.

[41] F. Gustafsson and F. Gunnarsson. Mobile positioning using wireless networks: possibilities and fundamental limitations based on available wireless network measurements. *IEEE Signal Processing Magazine*, 22(4):41–53, 2005.

[42] U. Hammes, C. L. Brown, R. F. Brcic, and A. M. Zoubir. Model selection for adaptive robust parameter estimation and its impact on multiuser detection. In *VIII IEEE Workshop on Signal Processing Advances in Wireless Communications (SPAWC)*, Helsinki, Finland, 2007.

[43] U. Hammes, E. Wolsztynski, and A. M. Zoubir. Semi-parametric geolocation estimation in NLOS environments. In *16th European Signal Processing Conference (EUSIPCO)*, Lausanne, Switzerland, August 2008.

[44] U. Hammes, E. Wolsztynski, and A. M. Zoubir. Transformation-based robust semiparametric estimation. *IEEE Signal Processing Letters*, 15:845–848, 2008.

[45] U. Hammes, E. Wolsztynski, and A. M. Zoubir. Robust tracking and geolocation for wireless networks in nlos environments. *IEEE Journal of Selected Topics in Signal Processing*, 3(5):889–901, October 2009.

[46] U. Hammes and A. M. Zoubir. A semi-parametric approach for robust multiuser detection in impulsive noise. In *IEEE International Conference on Acoustics, Speech, and Signal Processing (ICASSP '08)*, Las Vegas, USA, April 2008.

[47] U. Hammes and A. M. Zoubir. Mutiple model estimator based on M-estimation for tracking a mobile terminal in mixed LOS/NLOS environments. *IEEE Transactions on Signal Processing*, 2009. to be submitted.

[48] U. Hammes and A. M. Zoubir. Robust mobile terminal tracking in NLOS environments based on data association. *IEEE Transactions on Signal Processing*, 2009. to be submitted.

[49] F. R. Hampel, E. M. Ronchetti, P. J. Rousseeuw, and W. A. Stahel. *Robust Statistics, The Approach Based on Influence Functions*. John Wiley & Sons, 1986.

[50] A. K. Han. A non-parametric analysis of transformations. *Journal of Econometrics*, 35:191–209, 1987.

[51] P. Huber and E. M. Ronchetti. *Robust Statistics*. John Wiley & Sons, 2nd edition, 2009.

[52] P. J. Huber. Robust estimation of location parameter. *Annals of Mathematical Statistics*, 35(1):73–101, 1963.

[53] J. M. Huerta and J. Vidal. Mobile tracking using UKF, time measures and LOS-NLOS expert knowledge. In *IEEE International Conference on Acoustics, Speech, and Signal Processing, (ICASSP '05)*, volume 4, pages 901–904, 2005.

[54] J. M. Huerta and J. Vidal. Los-nlos situation tracking for positioning systems. In J. Vidal, editor, *IEEE 7th Workshop on Signal Processing Advances in Wireless Communications, (SPAWC '06)*, pages 1–5, 2006.

[55] Jose M. Huerta, Audrey Giremus, Josep Vidal, and Jean-Yves Tourneret. Joint particle filter and ukf position tracking under strong NLOS situation. In *14th IEEE Workshop on Statistical Signal Processing*, pages 537–541, 2007.

[56] S. S. Iyengar and R. R. Brooks. *Distributed Sensor Networks*. Chapman & Hall/Crc Computer and Information Science, 2004.

[57] J. A. John and N. R. Draper. An alternative family of transformations. *Applied Statistics*, 29:190–197, 1980.

[58] Simon J. Julier and Jeffrey K. Uhlmann. A new extension of the Kalman filter to nonlinear systems. pages 182–193, 1997.

[59] S. A. Kassam and H. V. Poor. Robust techniques for signal processing: A survey. *Proceedings of the IEEE*, 73(3):433–481, 1985.

[60] S. Kay. *Fundamentals of Statistical Signal Processing: Estimation Theory*. Prentice Hall, 1993.

[61] Kiseon Kim and G. Shevlyakov. Why Gaussianity? *IEEE Signal Processing Magazine*, 25(2):102–113, 2008.

[62] T. Kirubarajan and Y. Bar-Shalom. Probabilistic data association techniques for target tracking in clutter. *Proceedings of the IEEE*, 92(3):536–557, 2004.

[63] H. Krim and M. Viberg. Two decades of array signal processing research: the parametric approach. *IEEE Signal Processing Magazine*, 13(4):67–94, 1996.

[64] K. Kroschel. *Statistische Informationstechnik*. Springer, 4th edition, 2004.

[65] M. Gabbouj L. Yin, R. Yang and Y. Neuvo. Weighted median filters: A tutorial. *IEEE Transactions on circuits and systems*, 43(3):157–192, March 1996.

[66] B. Li, Y. Wang, H.K. Lee, A. Dempster, and C. Rizos. Method for yielding a database of location fingerprints in WLAN. *IEE Proceedings of Communications*, 152(5):580–586, 2005.

[67] J.-F. Liao and B.-S. Chen. Robust mobile location estimator with NLOS mitigation using interacting multiple model algorithm. *IEEE Transactions on Wireless Communications*, 5(11):3002–3006, 2006.

[68] R. A. Maronna, D. R. Martin, and V. J. Yohai. *Robust Statistics: Theory and Methods*. John Wiley & Sons Inc., 2006.

[69] C. Masreliez. Approximate non-Gaussian filtering with linear state and observation relations. *IEEE Transactions on Automatic Control*, 20:107–110, 1975.

[70] C. Masreliez and R. Martin. Robust bayesian estimation for the linear model and robustifying the Kalman filter. *IEEE Transactions on Automatic Control*, 22(3):361–371, 1977.

[71] E. Mazor, A. Averbuch, Y. Bar-Shalom, and J. Dayan. Interacting multiple model methods in target tracking: a survey. *IEEE Transactions on Aerospace and Electronic Systems*, 34(1):103–123, 1998.

[72] M. McGuire, K.N. Plataniotis, and A.N. Venetsanopoulos. Robust estimation of mobile terminal position. *Electronics Letters*, 36(16):1426–1428, 2000.

[73] M. McGuire, K.N. Plataniotis, and A.N. Venetsanopoulos. Data fusion of power and time measurements for mobile terminal location. *IEEE Transactions on Mobile Computing*, 4(2):142–153, 2005.

[74] D. Megnet, A. Rempfler, and H. Mathis. Autonomous combined GPS/GSM positioning. *European Journal of Navigation*, 2006.

[75] R. J. Meinhold and N. D. Singpurwalla. Robustification of Kalman filter models. *Journal of the American Statistical Association*, 84(406):479–486, 1989.

[76] P. Misra and P. Enge. *Global Positioning System Signals, Measurements and Performance*. Ganga-Jamuna Press, 2006.

[77] M. Najar, J. M. Huerta, J. Vidal, and J. A. Castro. Mobile location with bias tracking in non-line-of-sight. In *IEEE International Conference on Acoustics, Speech, and Signal Processing (ICASSP '04)*, volume 3, pages 956–959, 2004.

[78] M. Najar and J. Vidal. Kalman tracking for mobile location in nlos situations. In *14th IEEE Proceedings on Personal, Indoor and Mobile Radio Communications, (PIMRC '03)*, volume 3, pages 2203–2207, 2003.

[79] A. Papoulis. *Probability, Random Variables, and Stochastic Processes*. McGraw-Hill, 4th edition, 2002.

[80] T. Perala and R. Piche. Robust extended Kalman filtering in hybrid positioning applications. In *4th Workshop on Positioning, Navigation and Communication,*, pages 55–63, 2007.

[81] Y. Qi. *Wireless Geolocation in a Non-line-of-sight Environment*. PhD thesis, Princeton University, November 2003.

[82] Y. Qi, H. Kobayashi, and H. Suda. Analysis of wireless geolocation in a non-line-of-sight environment. *IEEE Transactions on Wireless Communications*, 5(3):672–681, 2006.

[83] J. Riba and A. Urruela. A non-line-of-sight mitigation technique based on ML-detection. In *IEEE International Conference on Acoustics, Speech, and Signal Processing (ICASSP '04)*, volume 2, pages 153–156, 2004.

[84] B. Ristic, S. Arulampalam, and N. Gordon. *Beyond the Kalman Filter: Particle Filters for Tracking Applications*. Artech House, 2004.

[85] X. Rong Li and V. P. Jilkov. Survey of maneuvering target tracking. part i. dynamic models. *IEEE Transactions on Aerospace and Electronic Systems*, 39(4):1333–1364, 2003.

[86] P. Rousseeuw and A. Leroy. *Robust Regression and Outlier Detection*. Wiley, 2003.

[87] P. Ruckdeschel. *Approaches for the robustification of Kalman filters (Ansätze zur Robustifizierung des Kalman-Filters)*. PhD thesis, Bayreuther Mathematische Schriften, 2001. vol. 64. In German.

[88] B. Sayadi and S. Marcos. A robustification method of the adaptive filtering algorithms in impulsive noise environments based on the likelihood ratio test. In *First International Symposium on Control, Communications and Signal Processing*, pages 685–688, 2004.

[89] A. H. Sayed, A. Tarighat, and N. Khajehnouri. Network-based wireless location: challenges faced in developing techniques for accurate wireless location information. *IEEE Signal Processing Magazine*, 22(4):24–40, 2005.

[90] I. C. Schick and S. K. Mitter. Robust recursive estimation in the presence of heavy-tailed observation noise. *Annals of Statistics*, 22:1045–1080, 1994.

[91] J. Schroeder, S. Galler, K. Kyamakya, and K. Jobmann. NLOS detection algorithms for ultra-wideband localization. In *4th Workshop on Positioning, Navigation and Communication, (WPNC '07)*, pages 159–166, 2007.

[92] M. I. Silventoinen and T. Rantalainen. Mobile station emergency locating in GSM. In *IEEE International Conference on Personal Wireless Communications*, pages 232–238, 1996.

[93] B. Silverman. *Density Estimation for Statistics and Data Analysis*. Chapman & Hall, 1986.

[94] B. Sinopoli, L. Schenato, M. Franceschetti, K. Poolla, M. I. Jordan, and S. S. Sastry. Kalman filtering with intermittent observations. *IEEE Transactions on Automatic Control*, 49(9):1453–1464, 2004.

[95] M. A. Spirito and A. G. Mattioli. Preliminary experimental results of a GSM mobile phones positioning system based on timing advance. In *In 50th IEEE Vehicular Technology Conference (VTC '99) - Fall*, volume 4, pages 2072–2076, 1999.

[96] G. Sun, J. Chen, W. Guo, and K. J. R. Liu. Signal processing techniques in network-aided positioning: a survey of state-of-the-art positioning designs. *IEEE Signal Processing Magazine*, 22(4):12–23, 2005.

[97] G. Sun and W. Guo. Bootstrapping M-estimators for reducing errors due to non-line-of-sight (NLOS) propagation. *IEEE Communications Letters*, 8(8):509–510, 2004.

[98] G. Sun and B. Hu. A minimum entropy estimation based mobile positioning algorithm. *IEEE Transactions on Wireless Communications*, 8(1):24–27, 2009.

[99] J.-A. Ting, E. Theodorou, and S. Schaal. A kalman filter for robust outlier detection. In *IEEE/RSJ International Conference on Intelligent Robots and Systems, (IROS '07)*, pages 1514–1519, 2007.

[100] D. J. Torrieri. Statistical theory of passive location systems. *IEEE Transactions on Aerospace and Electronic Systems*, AES-20(2):183–198, 1984.

[101] S. Venkatraman and J. Caffery. Statistical approach to non-line-of-sight BS identification. In *The 5th International Symposium on Wireless Personal Multimedia Communications*, volume 1, pages 296–300, 2002.

[102] M. Vossiek, L. Wiebking, P. Gulden, J. Wieghardt, C. Hoffmann, and P. Heide. Wireless local positioning. *IEEE Microwave Magazine*, 4(4):77–86, 2003.

[103] E. A. Wan and R. Van Der Merwe. The unscented kalman filter for nonlinear estimation. In *The IEEE Adaptive Systems for Signal Processing, Communications, and Control Symposium (AS-SPCC '00)*, pages 153–158, 2000.

[104] M. P. Wand, J. S. Marron, and D. Ruppert. Transformations in density estimation. *Journal of the American Statistical Association*, 86(414):343–353, June 1991.

[105] X. Wang and H. Poor. Robust multiuser detection in non-Gaussian channels. *IEEE Transactions on Signal Processing*, 47(2):289–304, February 1999.

[106] X. Wang, Z. Wang, and B. O'Dea. A TOA-based location algorithm reducing the errors due to non-line-of-sight (NLOS) propagation. *IEEE Transactions on Vehicular Technology*, 52(1):112–116, 2003.

[107] Y. Wang, J. Li, and P. Stoica. *Spectral Analysis of Signals, The missing data case*. Morgan & Claypool, 2005.

[108] A. J. Weiss. Direct position determination of narrowband radio frequency transmitters. *IEEE Signal Processing Letters*, 11(5):513–516, 2004.

[109] M. S. Williams. A regression technique accounting for heteroscedastic and asymmetric errors. *Journal of Agricultural, Biological, and Environmental Statistics*, 2:108–129, 1997.

[110] E. Wolsztynski, E. Thierry, and L. Pronzato. Minimum-entropy estimation in semi-parametric models. *Signal Processing*, 85(5):937–949, 2005.

[111] M. P. Wylie and J. Holtzman. The non-line-of-sight problem in mobile location estimation. In *5th IEEE International Conference on Universal Personal Communications*, volume 2, pages 827–831, 1996.

[112] I.-K. Yeo and R. A. Johnson. A new family of power transformations to improve normality or symmetry. *Biometrika*, 87:954–959, 2000.

[113] A. M. Zoubir and R. Brcich. Multiuser detection in heavy tailed noise. In *Digital Signal Processing: A Review Journal*, volume 12(2-3), pages 262–73, April-July 2002.

[114] A. M. Zoubir and D. R. Iskander. *Bootstrap Techniques for Signal Processing*. Cambridge, 2004.

Publications

Internationally Refereed Journal Articles

- U. Hammes and A. M. Zoubir. Mutiple Model Estimator Based on M-estimation for Tracking a Mobile Terminal in mixed LOS/NLOS Environments. *IEEE Transactions on Signal Processing*, 2010, under review.

- U. Hammes and A. M. Zoubir. Robust Mobile Terminal Tracking in NLOS Environments Based on Data Association. *IEEE Transactions on Signal Processing*, 2009, under review.

- U. Hammes and E. Wolsztynski and A. M. Zoubir. Robust Tracking and Geolocation for Wireless Networks in NLOS Environments. *IEEE Journal of Selected Topics in Signal Processing*, vol. 3, no. 5, pp. 889-901, October 2009.

- U. Hammes, E. Wolsztynski and A. M. Zoubir. Transformation-Based Robust Semi-Parametric Estimation. *IEEE Signal Processing Letters*, vol. 15, pp. 845-848, December 2008.

Internationally Refereed Conference Papers

- M. Muma, U. Hammes, A. M. Zoubir, Robust Semiparametric Amplitude Estimation of Sinusoidal Signals: The Multi-Sensor Case, *In Proceedings of the Third International Workshop on Computational Advances in Multi-Sensor Adaptive Processing (CAMSAP)*, Aruba, Dutch Antilles, December 2009.

- C. Fritsche, U. Hammes, A. Klein and A. M. Zoubir. Robust Mobile Terminal Tracking in NLOS environments using Interacting Multiple Model Algorithm. *The 34th IEEE International Conference on Acoustics, Speech and Signal Processing (ICASSP)*, Taipei, Taiwan, April 2009.

- U. Hammes, E. Wolsztynski and A. M. Zoubir. Semi-parametric Geolocation Estimation in NLOS Environments. *16th European Signal Processing Conference (EUSIPCO)*, Lausanne, Switzerland, August 2008.

- U. Hammes, A. M. Zoubir. A Semi-parametric Approach for Robust Multiuser Detection in Impulsive Noise. *In Proceedings of the 33rd IEEE International Conference on Acoustics, Speech and Signal Processing (ICASSP)*, Las Vegas NV, USA, March 2008.

- U. Hammes and C. L. Brown and R. F. Brcic and A. M. Zoubir. Model Selection for Adaptive Robust Parameter Estimation and its Impact on Multiuser Detection. *In VIII IEEE Workshop on Signal Processing Advances in Wireless Communications (SPAWC)*, Helsinki, Finland, June 2007.

- U. Hammes, R. F. Brcic, A. M. Zoubir. A Robust Transmitter Diversity Scheme for CDMA in Impulsive Noise. *In International ITG/IEEE Workshop on Smart Antennas, (WSA 2007)*, Vienna, Austria, February 2007.

I want morebooks!

Buy your books fast and straightforward online - at one of world's fastest growing online book stores! Environmentally sound due to Print-on-Demand technologies.

Buy your books online at
www.morebooks.shop

Kaufen Sie Ihre Bücher schnell und unkompliziert online – auf einer der am schnellsten wachsenden Buchhandelsplattformen weltweit! Dank Print-On-Demand umwelt- und ressourcenschonend produziert.

Bücher schneller online kaufen
www.morebooks.shop

KS OmniScriptum Publishing
Brivibas gatve 197
LV-1039 Riga, Latvia
Telefax: +371 686 204 55

info@omniscriptum.com
www.omniscriptum.com

Printed by Books on Demand GmbH, Norderstedt / Germany